AF345272

Calcium, Cadherins, and Cell Adhesion: A Three-Way Dance

Lemony

Copyright © 2024 by Lemony

All rights reserved. No part of this book may be reproduced in any man-ner whatsoever without written permission except in the case of brief quotations embodied in critical articles and reviews.
First Printing, 2024

Table of Contents

Chapter 1. Introduction

Cell-cell Adhesion

Complex multicellular organisms are composed of many distinct but interconnected systems of organs and tissues in order to perform the functions necessary for life. These systems must coordinate for the organism to move, to digest the food they take in, to breakdown toxins, and to perceive and interact with the world around them. Communication between different cells is necessary to ensure that each piece of the system works harmoniously and changes or interference to that communication can lead to often deleterious consequences for the organism. One method of communication is through cell adhesion, using molecules attached to the surface of cells interacting with one another. Cellular adhesion plays significant roles in embryogenesis, tumor suppression, tissue morphogenesis, immunological response, and mechanosensation. Proteins that form adhesive interactions between cells are called cell adhesion molecules, or CAMs.

There are several families of CAMs expressed in vertebrates: cadherins[1], selectins[2], integrins[3] and immunoglobulins[4]. While each group has distinct functions and use different mechanisms to interact with other molecules, the CAMs have one overall function in common: connect the outside of the cell to the inside.

The Cadherin Superfamily

The cadherins are a diverse family of calcium dependent cell adhesion proteins. Since the discovery of E-cadherin[5,6] and its later structural characterization, over 100 unique members have been identified in the human genome alone (Figure 1A)[1,7,8]. Each member of the family carries out a specific role in the body, regulating cell development and embryogenesis[1,8], guiding assembly of neural circuitry[9–11], and mediating mechanosensation[12–16]. To carry out these biological roles, cadherins have evolved a well-defined domain architecture suited to connect the outside of the cells to the inside. On the extracellular side, each cadherin contains a set of tandem extracellular cadherin (EC) repeats. The extracellular domain is typically connected to the cell membrane through a transmembrane helix, which is followed by a cytoplasmic domain that often interacts with other proteins involved in connections to the cytoskeleton or cell signaling[17].

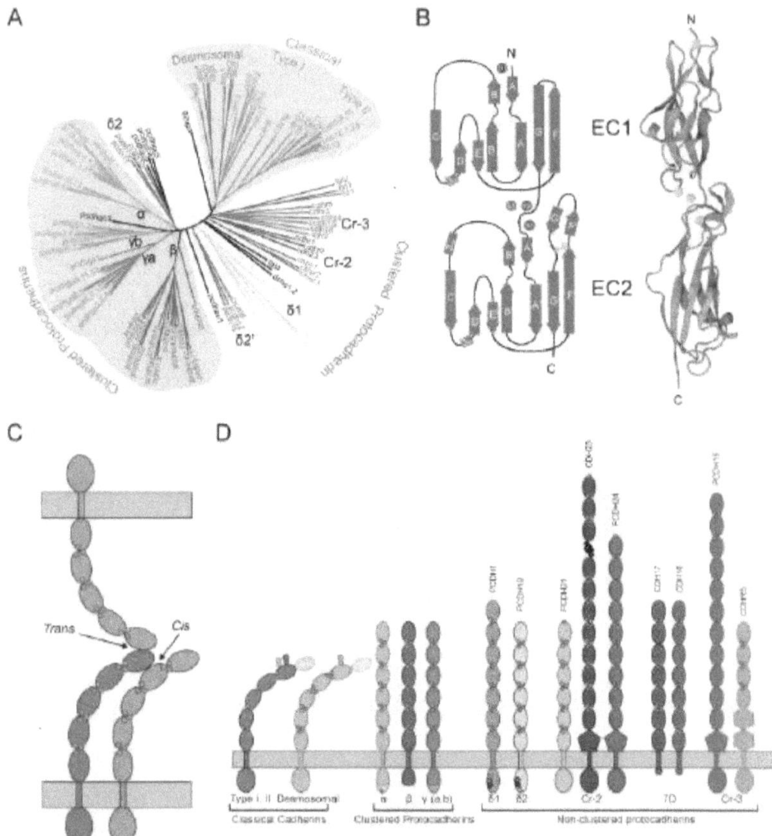

Figure 1.1 The cadherin superfamily and domain architecture

A) A family tree of all cadherins within the human genome ordered into subfamilies and groups by the sequence alignment of EC1-3 (adapted from Sotomayor et. al. 2014). Three distinct subfamilies exist: classical cadherins, clustered protocadherins and non-clustered protocadherins. B) Topology diagram and cartoon representation of an EC1-2 fragment of PCDH24 (adapted from Chapter 3). The β-strands are denoted A-G, forming a Greek-key motif which can be seen in the cartoon and topology schematic. Three calcium ions are bound by residues at the linker region between repeats and are shown here as green spheres. C) Schematic of *trans* and *cis* interactions using a classical cadherin domain architecture (adapted from Sotomayor et. al. 2014). Interactions in *trans* come from opposing cell membranes while interactions in *cis* involved two monomers on the same cellular membrane. D) Domain architecture of a variety of cadherins. Ovals denote EC repeats, and green spheres denote calcium ions. Pentagons denote membrane adjacent domains (MAD) in non-clustered protocadherins and hexagons denote mucin-like domains. Various shapes and domains are used to represent the cytoplasmic domains and their conserved motifs.

The EC repeats make up the adhesive unit of a cadherin, some of them having as few as two EC repeats and others having greater than 30. Each EC repeat is composed of around 100 amino acids arranged in a 7 β-strand Greek key motif (Figure 1.1B)[18]. Within the sequence of EC repeat are conserved calcium binding motifs, which allow a pair of EC repeats to bind up to three calcium ions at the linker region between repeats (Figure 1.1B). The binding of calcium is responsible for imparting rigidity on the extracellular domain[19–21] and disruptions caused by mutations in calcium-binding residues often results in deleterious consequences for adhesion[13,22]. The EC repeats have the same overall fold and arrangement; however they are not identical within a cadherin and some cadherins contain EC repeats that lack calcium binding residues, leading to flexibility[23–26]. The differences in the number of EC repeats (Figure 1.1D), and their sequences allow cadherins to have a wide variety of adhesive interfaces and cellular functions[8,10,22,27–33].

Cadherins mediate adhesion through their EC repeats in two ways. In order to directly link two adjacent cellular membranes, the EC repeats overlap in a *trans* configuration (Figure 1.1C). These interactions are typically homophilic, i.e. among cadherins of the same type. However exceptions exist for one type of cadherin to interact with another, in a heterophilic manner. Alternative to the *trans* configuration, the extracellular domains of two cadherins can come from the same membrane, interacting in a *cis* fashion to add additional strength or specificity (Figure 1.1C)[27,34]. Most cadherins use a combination of *trans* and *cis* interactions to mediate adhesion. Due to the variety of function, domain architectures, and adhesive mechanisms, the superfamily can be broken down into subfamilies (Figure 1.1A, D): the classical cadherins (type I, II, desmosomal cadherins), clustered protocadherins (α, β, and γ), and non-clustered protocadherins (Cr-1, Cr-2, δ1, and δ2).

In this chapter, I will introduce each of the subfamilies in more detail, highlighting how each subfamily mediates cell adhesion. I will begin with the classical cadherins, the most understood subfamily, followed by the clustered protocadherins and finally introducing the unique interactions seen in the non-clustered protocadherins. I will then introduce in more detail the specific cadherins I have studied.

Classical Cadherins

Classical cadherins make up the most studied and understood subfamily of cadherins. They are ubiquitously expressed in a wide variety of cell types, supporting the cells typically subjected to mechanical stress and epithelial barriers. They also play an important role in embryogenesis and morphogenesis. All classical cadherins have five EC repeats (Figure 1.1D, 1.2A), a transmembrane domain, and an extended cytoplasmic domain that interacts with cytoplasmic proteins to bind to the cell's cytoskeleton [8,17]. These cadherins also have a prodomain N-terminal to EC1 that must be cleaved for them to form adhesive complexes[35]. Members of the subfamily interact in nearly the same fashion, but belong to one of three types: type I, type II, and desmosomal cadherins.

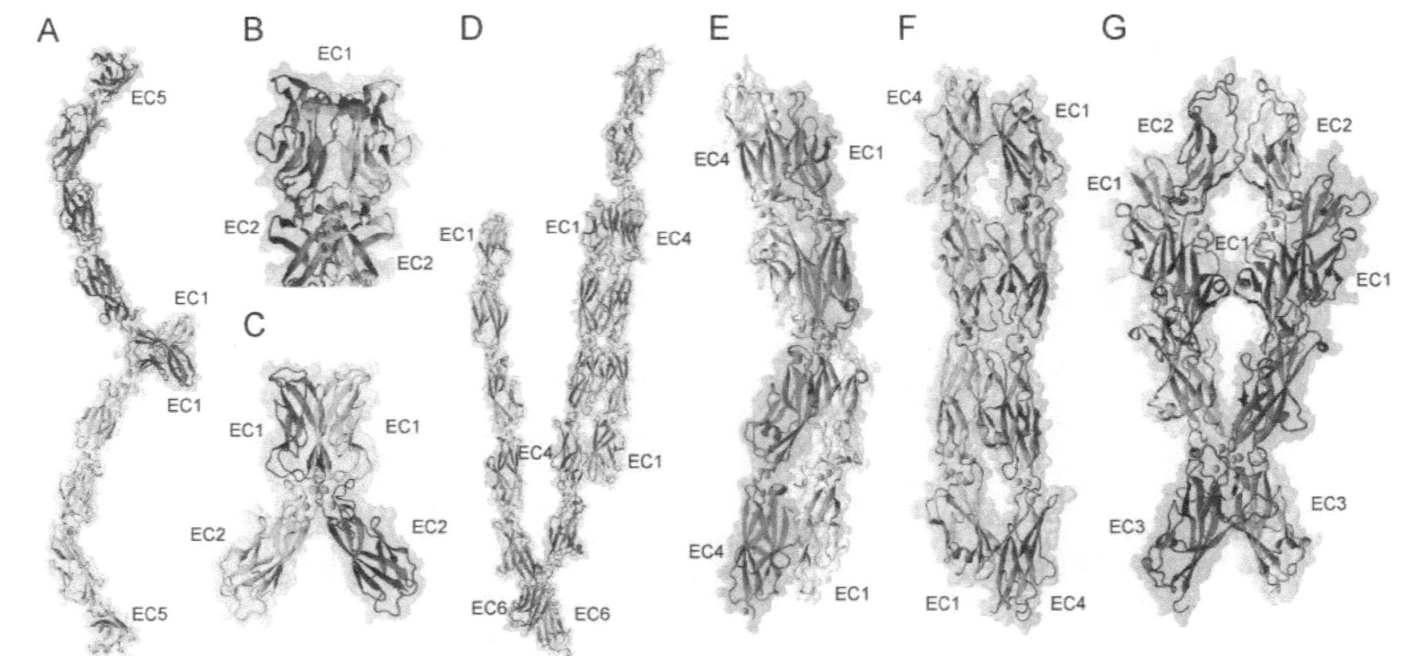

Figure 1.2 Adhesion mechanisms of various members of the cadherin superfamily

A-G) Ribbon representation overlaid onto surface representation of various cadherin structure engaged in *trans* and *cis* interactions. These adhesive mechanisms have been determined from crystal structures of cadherins in the classical cadherins, clustered protocadherins and non-clustered protocadherins subfamilies. A) The *trans* interaction of CDH1, including EC1-5. The *trans* interaction occurs between EC1 repeats[34]. B) Close up of the tryptophan (W2) strand-swap observed in CDH1 for classical type I cadherins. C) The *trans* EC1-2 of T-cadherin in the X-dimer conformation[28]. D) The *trans* and *cis* interactions of PCDH γB4 EC1-6. The *trans* interaction includes repeats EC1-4 while the *cis* interaction involves the EC6 repeat [27]. E-F) The *trans* interaction observed for δ1 and δ2 protocadherins. The interaction takes place between EC1-4 for both PCDH1 (E)[31] and PCDH19 (F)[22]. G) The heterotetramer of the inner ear tip-link protocadherins, CDH23 (in blue) and PCDH15 (in purple)[36,37]. The handshake interaction occurs in *trans* between CDH23 EC1-2 and PCDH15 EC1-2. A *cis* interaction between PCDH15 EC2-3 can be seen in this structure.

Type I classical cadherins, like E-cadherin (CDH1) and N-cadherin (CDH2), exchange a conserved tryptophan (W2) residue in the most N-terminal EC repeat, EC1, with another EC1 to form a strand-swap interaction in *trans* (Figure 1.2A-B)[8,38]. For the strand-swap to occur, the EC1-2 domains go through an intermediate state, known as the X-dimer (Figure 1.2C) where the calcium-binding linkers overlap[28,39–41]. This intermediate conformation is thought to bring the EC1 domains closer, thereby allowing the N-terminal W2 residues to unbind from their own hydrophobic pockets and insert into the opposing EC1 monomer. Interestingly, this X-dimer conformation is the primary mode of adhesion for a unique classical cadherin member, T-cadherin (CDH13), which amongst other unusual sequence features, lacks the W2 residue needed to strand-swap (Figure 1.2C)[28].

The strand-swap *trans* interaction is further strengthened by a *cis* interaction between the EC1 repeat of one protomer and the EC2 repeat of another monomer on the surface of the same cell. In this way, type I classical cadherins can form Velcro-like patches on the cell surface[34,42]. Type II classical cadherins act in a similar fashion, but have two conserved tryptophans (W2, W4), both of which are exchanged between EC1 monomers to form the strand swap[43]. Desmosomal cadherins have just one conserved tryptophan (W2) in the same position as Type I classical cadherins, but the two types of desmosomal cadherin, desmoglein and desmocollin, are capable of forming heterophilic complexes[30] whereas type I and type II classical cadherins are mostly homophilic, though some exceptions exist[44,45].

Clustered Protocadherins

Over half of the cadherins in the human genome are clustered protocadherins (Figure 1.1A), which are expressed in the neurons, where they guide the formation of neuronal synaptic connections. Each neuron expresses a small subset of clustered protocadherins, effectively giving it a unique "barcode" by which to identify and avoid making connections with itself. This mechanism of identification is termed neuronal self-avoidance[46,47]. Clustered protocadherins are able to facilitate self-avoidance by encoding splice variants in three genes clusters α, β, and γ, similar to immunoglobulins[9,10,27,29,48–51]. Each gene cluster can produce a wide variety of protocadherins with unique sets of six EC repeats.

It was originally assumed that all cadherins mediate *trans* adhesion via one or two N-terminal repeats, however experiments revealed that clustered protocadherins mediate *trans* interactions with EC1-4 and in a homophilic fashion (Figure 1.2D)[29,32]. The *cis* interactions mediated by clustered protocadherins, however, are heterophilic and occur between the most C-terminal repeat, EC6 (Figure 1.2D)[27,48]. When the specific clustered protocadherins expressed in a neuron come together, the heterophilic *cis* interactions form the unique barcode mentioned above. If that set of clustered protocadherins then interact in *trans* with all the same variations on an opposing membrane, the neuron can recognize that it is interacting with itself and avoids forming a synaptic interaction with that membrane. However, in the absence of any or all *trans* interactions, the neuron can recognize the opposing membrane as a different cell and forms a connection to it[52].

Non-clustered Protocadherins

The most diverse and unique group of cadherins is formed by the members of the subfamily of non-clustered protocadherins. Members in this subfamily are involved in processes that range from neuronal recognition, planar cell polarity, to sound and vision perception, as well epithelia morphogenesis and maintenance in a variety of organs[7,13]. The subfamily can be broken down in various groups based on their distinct modes of adhesion. Unlike the classical and clustered protocadherins, the length and number of EC repeats within the non-clustered protocadherins is highly variable. Some groups have a conserved number of repeats, such as the two δ-protocadherin groups. Protocadherins within the δ1 group have seven EC repeats, while the δ2 protocadherins have six (Figure 1.1D)[53,54]. Other groups have similar modes of binding but not a conserved number of EC repeats. This is the case for both the Cr-2 and Cr-3 groups. Within the Cr-2 group, there is cadherin 23 (CDH23) with 27 EC repeats, protocadherin 21 (PCDH21, CDHR1) with six repeats, and protocadherin 24 (PCDH24, CDHR2) with nine EC repeats (Figure 1.1D)[55,56]. Similarly, in the Cr-3 group, protocadherin 15 (PCDH15) has 11 EC repeats[57–60], while mucin-like protocadherin (CDHR5) has only four EC repeats (Figure 1.1D)[61]. Others such as the FAT cadherins, and Daschous (DCHS) cadherins have very long extracellular domains, with 34 and 27 EC repeats, respectively[62].

The two groups of δ protocadherins are excluded from the clustered protocadherins because these are not on clustered genes and are not splice variants. In addition to extracellular domains mentioned above, the δ protocadherins have a transmembrane domain linking them to a cytoplasmic domain that has conserved sequence motifs. These protocadherins are expressed in a wide variety of cell types and are associated with neurological diseases, asthma, and viral infections[63,64]. Similar to the clustered protocadherins, δ protocadherins mediate *trans* adhesion via EC1-4, as was discovered for δ1 member protocadherin 1 (PCDH1) (Figure 1.2E) and δ2 member protocadherin 19 (PCDH19) (Figure 1.2F)[22,31]. Interestingly, bead aggregation assays revealed that PCDH19 can form *cis* interactions with the classical cadherin, CDH2, with this *cis* heterodimer forming a *trans* homodimer, giving PCDH19 two different modes of cell adhesion[45]. However, the structural basis for the *cis* interactions in δ-protocadherins is unclear.

The Cr-2 and Cr-3 families are known to mediate heterophilic *trans* between each other. The first structural model of such a heterophilic interaction was found in the inner ear tip-link cadherins, CDH23 (a member of Cr-2) and PCDH15 (a member of Cr-3). Both are long cadherins, essential for forming the filamentous tip link that allows for sound mechanotransduction[57,65] and their heterophilic *trans* interaction is limited to the first two EC repeats of each cadherin respectively (Figure 1.2G)[33]. The overlap is dubbed the "handshake" interaction and was validated by the presence of essential residues in the interface that lead to deafness when mutated[65,66]. Based on microscopy images of the tip link as well as electron microscopy data, there are at least two molecules of CDH23 and PCDH15 forming the tip link, making the tip link a heterotetramer (Figure 1.2G). This was later supported by a crystal structure showing a heterotetramer formed by two molecules of CDH23 EC1-2 and two molecules of PCDH15 EC1-3 (Figure 1.2G) where the heterophilic handshake was observed, along with a *cis* interaction present between PCDH15 EC2-3 and also seen in the structure of PCDH15 EC1-3[37]. Later a second pair of

Cr-2/Cr-3 family members were discovered to interact in a heterophilic fashion: PCDH24 and CDHR5, of which I will go into greater detail later in this chapter[67].

Despite several of the non-clustered protocadherins now being studied and characterized, the adhesion and function of many in this diverse subfamily have not been elucidated. This book focuses on several non-clustered protocadherins, including the 7D-cadherins, cadherin 17 (CDH17, also known as LI-cadherin, BILL-cadherin) and cadherin 16 (CDH16, also known as Ksp-cadherin), the Cr-2/Cr-3 pair involved in the intestinal brush border, PCDH24 and CDHR5, and another Cr-2 family member, PCDH21. In the sections that follow, I will go into greater detail about these specific cadherins.

Cadherins of the Gut Epithelia

The digestive system is composed of several different organs that function to digest food, absorb necessary nutrients from that food, and expel any waste leftover. It provides an ecosystem for a diverse array of bacteria that aid in digestion, which it must keep in balance and therefore is also considered a part of the immune system. The organ that absorbs most of the nutrients from digested food is the small intestine, and evolution has developed a clever way of ensuring that a large surface area can fit into the mammalian body, regardless of size. The small intestine is made up of intestinal folds, called villi (Figure 1.3A), which increase the inside surface area. The villi are lined with specialized columnar epithelial cells called enterocytes and their apical sides are coated with thousands of tiny actin-filled protrusions called microvilli (Figure 1.3B-C)[68]. The combination of the villi and microvilli form the intestinal brush border and increase the surface area by 1000-

fold. Additionally, the small intestine is subject to mechanical forces, as material is moved down the digestive tract. The enterocytes express several types of cadherins to properly function and to maintain the brush border.

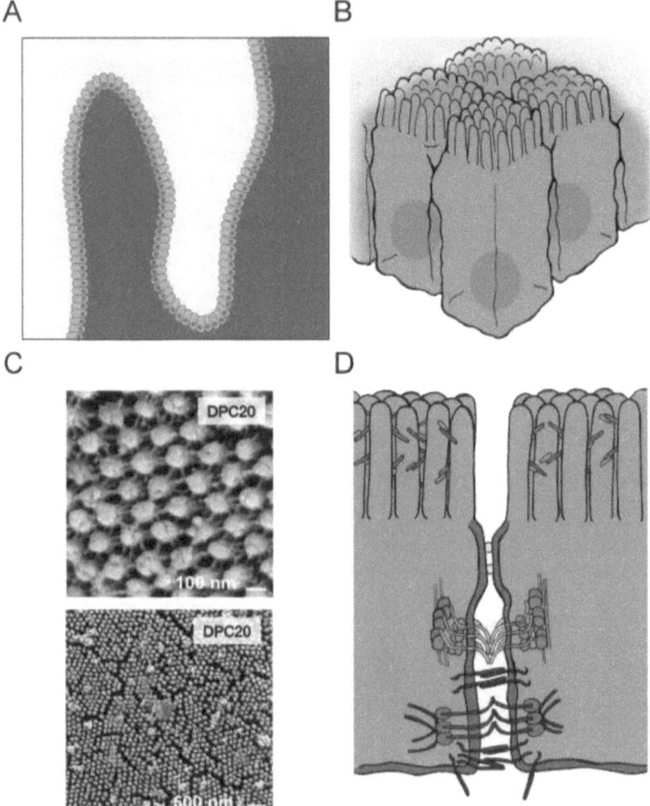

Figure 1.3 Morphology and cadherins of the intestine and location of gut cadherins.
A) Schematic of the villi that line the small intestine. The intestinal lumen is denoted in yellow, with the villi in purple forming folds into the lumen. The villi are coated with a layer of specialized columnar epithelial cells, called enterocytes (grey with red cell nuclei). B) Schematic of the shape of intestinal enterocytes. The cell body is polarized, with the nucleus (red) localized near the basolateral side. The apical or top side is lined with actin-filled microvilli, which increase the surface area of the enterocyte and form the brush border. C) Electron micrographs of the brush border in CACO-2 cells (adapted from Crawley et. al. 2014). This view is from above and shows the links between microvilli (top) and the hexagonal arrangement of the microvilli (bottom). D) Schematic representation of localization of various cadherin family members within the intestinal epithelia. The adherens junction is represented by CDH1 in light blue. The desmosomes, composed of desmocollin and desmoglein (both magenta) are shown below. CDH17 (navy) is distributed along the basolateral membrane, excluded from the adherens junctions and desmosomes. Finally, the intermicrovillar links composed of PCDH24 (red) and CDHR5 (orange) are shown linking adjacent microvilli together on the apical side.

The brush border enterocytes are protected from shearing by mechanical forces by the classical cadherin CDH1 and the desmosomal cadherins. CDH1, using its adhesive interactions in combination with proteins in the cytoplasm, forms strong links between the outside of the cell and the actin-cytoskeleton inside. These complexes are localized to the top third of the lateral membrane and are part of the adherens junction[17,34,69]. In a similar fashion, the desmosomal cadherins form the desmosomes in the lower lateral membrane, linking instead to the intermediate filaments[30,70].

Excluded from the adherens junctions and the desmosomes is the non-classical CDH17, which is likely to be involved in the regulation of water transport between the epithelial cells[71-73]. It is localized along the lateral and basal portions of the epithelia. On the apical side, the microvilli are formed and maintained by two non-clustered protocadherins, PCDH24 and CDHR5, together known as the intermicrovillar links[67]. All three of these cadherins have been the focus of my research and have related family members, CDH16 (related to CDH17) and PCDH21 (related to PCDH24).

CDH17 and the 7D-cadherins

CDH17, and its sibling, CDH16 form the small 7D-cadherin subfamily, named because both proteins have seven EC repeats instead of the five observed in classical cadherins. The 7D-cadherins lack the prodomain as well as the tryptophan in EC1 necessary to mediate the strand-swap mechanism, and also have a short cytoplasmic domain (~20-30 residues) in contrast to the long cytoplasmic domains (~150 residues) seen in classical cadherins[74,75].

Gene analysis revealed that in addition to sequence similarities between CDH17 and CDH16, the gene structure of exons and introns is also remarkably similar, suggesting that CDH17 and CDH16 have a common ancestor[76,77]. Additionally, the EC1-2 and EC3-4 repeats share sequence similarity between one another, as well as with repeats EC1-2 of classical cadherins, revealing that repeats EC1-2 of the 7D-cadherins likely arose from a gene duplication event from a classical-like ancestor[76]. The EC3 repeats of both CDH17 and CDH16 have a tryptophan residue on the N-terminus, which could have been the strand-swap tryptophan present in the classical-like ancestor[76]. The gene duplication hypothesis is further supported by sequence analysis of the EC2 domains of the 7D-cadherins, which lack canonical residues necessary for binding calcium ions at the EC2-3 linker. This evidence suggests that the EC2-3 repeats might lack some, or all, of the calcium ions in its linker region. The remainder of repeats, EC5-7, also share similarities to EC3-5 in classical cadherins[76,77].

CDH17 and CDH16 are similar in their domain architecture and are expressed in a wide variety of systems. Little is known about how these proteins mediate adhesion. Although, more is known about CDH17 than CDH16, as detailed below.

CDH17

CDH17 was originally identified in the human intestine, and rat liver[73,74]. Studies have looked at its expression in mouse intestines[78], in the spleen and B-cells[79–81] as well as in the kidneys of zebrafish[82]. Expression of CDH17 in mice occurs only in epithelial cells lining the intestinal lumen, where it was present in cells of both the crypts and along the villi[78]. However, Northern blot analysis, which measures the expression of RNA, revealed that mRNA for CDH17 is also produced in the spleen. It was observed that CDH17 mRNA was produced during development but later than mRNA for CDH1, which is also expressed in intestinal epithelia[78]. The function of CDH17 in the intestinal epithelia is currently unknown, however there is evidence to support that it plays a role in the regulation of water transport between intestinal epithelial cells[71,72]. The interstitial cleft (IC) width can be changed depending on the electrolyte concentration, which in turn affects how CDH17 mediates adhesion as the *trans* interaction of CDH17 is dependent on calcium concentration[71].

The function of CDH17 in the spleen and in mature B cells in the bone marrow is different than in the intestinal epithelia[79–81]. CDH17 expression in B cells was found to be important during various stages of B cell development, as demonstrated by mice deficient in CDH17 that showed less development of immature B cells compared to WT[80]. Additional work showed that CDH17 is important for the survival and sorting of memory B cells, B cells that have been attuned to specific antigens and stored in case of later infection[81]. It was additionally revealed that the spleen, where CDH17 is expressed, provides one niche for the survival of memory B cells, which is dependent on homophilic adhesion of CDH17[81]. A compound heterozygous mutation on CDH17 in one individual, where each copy of the gene carry a different mutation, has been identified to lead to severe immunodeficiency[83].

In addition to roles in the development of the intestinal epithelia, water transport, and B-cell maturation, CDH17 has been implicated in a variety of gastric, pancreatic, colorectal, and hepatocellular cancers[84]. Misregulation of CDH17 leads to a higher rate of metastasis in cancer cells, as well as metaplastic changes in mucosal cells, such as those in Barrett's esophagus, that express features more commonly seen in the intestinal epithelia[85,86]. Understanding how CDH17 mediates adhesion would be useful in designing drug targets to treat cancers where its expression is observed, given its expression in relatively few locations in the body.

Cadherin 17 is capable of mediating *trans* and *cis* interactions between cells[44,72,87–89]. Early work done on CDH17 revealed that CDH17 does not co-purify with β-catenin, a cytoplasmic binding partner of classical cadherins and that it does not need cytoplasmic partners to mediate adhesion, as a GPI-anchored construct of CDH17 was able to mediate adhesion [87]. However, its *trans* adhesion is highly dependent on calcium concentration and has a lower affinity for calcium than E-cadherin[88]. This sensitivity is suggested to affect the IC width dependent on the electrolyte concentration present in the intestinal epithelia[71]. In experiments that determined CDH17's involvement in IC width maintenance, a short peptide that interfered with *trans* adhesion was developed[72]. This peptide (VAALD) was made based on a crystallographic model of the EC1 of a classical cadherin. Interestingly, CDH17 is capable of mediating heterophilic *trans* interactions with E-cadherin, with a similar affinity as its homophilic *trans* interaction[44]. Whether this interaction is physiological or not remains unknown, as CDH17 localization is excluded from the adherens junction where E-cadherin is present. Additionally, CDH17 can form homophilic *cis* interactions, as inferred by fluorescence resonance energy transfer (FRET) experiments[89]. A donor-acceptor pair of fluorophores were attached to the C-terminal ends of CDH17 and expressed in HEK293 cells, when the donor tag was excited, high FRET efficiency for emission of the acceptor tag was observed. Even when depletion of calcium by EGTA reduced the *trans* interaction, CDH17 dispersed across the cellular membrane and FRET efficiency was not affected, suggesting that the CHD17 *cis* interaction is a calcium-independent interaction.

CDH16

CDH16 was the second 7D-cadherin to be discovered and it was sequenced from rabbit, mouse and human kidney[75,90] and later found to be expressed in the human thyroid[91,92]. Comparisons of its sequence and domain organization quickly suggested it was related to CDH17[93]. However, less is known about CDH16 and its function is unclear. It is capable of mediating *trans* adhesion, like CDH17, and it does not rely on interactions with cytoplasmic partners to mediate adhesion[93]. Unlike CDH17, very little research exists outside of gene and sequencing analyses and its involvement in cancer.

CDH16 has been linked to cancers of the kidney[94–99] and thyroid [91]. Changes in expression of CDH16 vary from different tumor types, which makes it ideal in identifying cancer type. Understanding the adhesive interactions of CDH16 and its relationship to CDH17 might allow for new cancer treatments.

PCDH24 and CDHR5

Two non-clustered protocadherins, PCDH24 and CDHR5, form the extracellular component of the intermicrovillar adhesion complex (IMAC)[67,100–104], a network of cytoplasmic and transmembrane proteins that are essential in the development and maintenance of the brush border epithelia, specifically the apical microvilli. PCDH24 and CDHR5 are required for the linking together of the microvilli in an organized hexagonal array[67,100]. Experiments with PCDH24 mutants, knockdowns and with PCDH24 knockout mice have demonstrated that elimination of the IMAC leads to remarkable disorganization of the microvilli[67,103]. The PCDH24 knockout mice, while viable, had lower body weight than the WT and had defects in the packing of microvilli in the brush border[103].

PCDH24 and CDHR5 are related to the inner-ear tip-link protocadherins, CDH23 and PCDH15, respectively. PCDH24 is shorter than CDH23, with nine repeats and a membrane adjacent domain (MAD)[105–108]. However, sequence alignments of its N-terminal repeats reveal that it has features similar to those observed for the N-terminus of CDH23[7,33,109,110]. The sequence of PCDH24 has residues that might be involved in an atypical calcium binding site on the most N-terminal repeat, EC1. This is considered a canonical feature of the Cr-2 group of non-clustered protocadherins[109]. Similarities between CDHR5 and PCDH15 are also observed. CDHR5 is much shorter than PCDH15, with only four EC repeats. However, the sequence of CDHR5 EC1 features a predicted disulfide bond near its tip, as observed in PCDH15. Additionally, a mutation in CDHR5, R84G, was shown to interfere with the organization of microvilli[33,67]. This mutation is similar to a deafness mutation in PCDH15, which breaks the handshake[111]. The similarities of PCDH24 and CDHR5 to CDH23 and PCDH15 suggest that these may use a handshake-like interaction to form and maintain the microvilli.

PCDH21

PCDH21 is a member of the Cr-2 group of non-clustered protocadherins and is related to CDH23 and PCDH24[109]. It was originally cloned along with CDH23[112] and is expressed in the rods and cones of the retinal epithelia[113]. It is the shortest of the Cr-2 group members, with six EC repeats, and currently it is unknown if PCDH21 interacts in a homophilic manner with itself. It has a unique cytoplasmic domain sequence, which is not similar to the cytoplasmic domain sequences of the Cr-2 family members, CDH23 and PCDH24, both of which interact with a similar group of cytoplasmic proteins to attach to the cytoskeleton[24,67,102,114]. PCDH21 has been shown to interact with prominin 1 (PROM1). These proteins co-localize to the base of the outer segment of the rod and cone photoreceptor cells. PCDH21 plays a role in the development and regeneration of the photoreceptor discs in rods and cones, and disruptions to its expression is detrimental to the formation of the discs and subsequently, vision[113,115].

Mutations in PCDH21 lead to retinitis pigmentosa, a progressive form of blindness[116–125]. Most of the mutations affect splicing or produce a nonsense mutation within the coding region of the protein. Although this clarifies how critical PCDH21 is for the maintenance of the photoreceptors, no missense mutations have been identified in the N-terminal EC repeats that may interfere with function and a possible *trans* interaction with itself or with other proteins. A missense mutation in EC6 (I676N) has been previously identified[122,125] and could potentially interfere with a *cis* interaction or other binding partners. A study in 2018 identified PCDH21 among six potential contributors to the development of bipolar disorder[126].

PCDH21, unlike its Cr-2 siblings, does not currently have a known heterophilic cadherin binding partner in the Cr-3 family, although its N-terminal repeats share the same unique site 0 for calcium binding in EC1. Currently, it is known to interact with PROM1, another gene involved in retinal disease. PROM1 is a glycoprotein, and its localization depends on PCDH21, as homozygous mouse knockouts of PCDH21 showed that PROM1 was unable to localize to the base of the outer segment (OS) and was instead spread out across the membrane. Despite what little is known about its adhesion or potential binding partners, understanding PCDH21 on a structural level may provide insight into how it functions in photoreceptors and tools to explore its role in disease.

book Outline

The non-clustered, non-classical cadherins CDH17, PCDH24, CDHR5, CDH16, and PCDH21 have been the focus of my work. In Chapter 2, I describe our results on the structure of the N-terminus of CDH17 and reveal what EC repeats are necessary for *trans* adhesion. In Chapter 3, I show the structures and interfaces of the intermicrovillar cadherins, PCDH24 and CDHR5. My initial work on PCDH21 and CDH16 is described in Appendices A and B, respectively.

Chapter 2. More than a handshake: How cadherin 17 adheres cell together1

Introduction

Cadherins form a large superfamily of calcium-dependent cell adhesion proteins, with around 100 structurally diverse members encoded in the human genome alone[7,8]. These proteins have an extracellular domain of variable length, a transmembrane domain, and a cytoplasmic domain that often facilitates interaction with the cytoskeleton and other signaling proteins[17]. The sequence and structural diversity of cadherins allow them to perform a wide variety of functions in development and morphogenesis[1,8,127], mechanosensation[12–16,128] and neuronal recognition[9,11,129–131]. The most studied and nearly ubiquitous member of the family, E-cadherin (CDH1), forms the adherens junction in epithelia, is essential in embryogenesis, and is considered a tumor suppressor[132]. Other cadherins have more specific functions. For instance, cadherin-23 and protocadherin-15 form a bond essential for sound mechanotransduction[33,65], while clustered protocadherins use their diverse extracellular domains to mediate neuronal recognition[27,32,51,133]. Less is known, however, about non-classical and non-clustered members of the superfamily.

Cadherin-16 (CDH16, aka Ksp-cadherin) and cadherin-17 (CDH17, aka LI-cadherin or BILL-cadherin) are the only members of the small subfamily of 7D-cadherins in vertebrates[73,75,77]. CDH16 is found in kidney epithelia[75,90], while CDH17 is expressed in human and mouse intestinal cells[73,78], rat liver[74], and in mature B cells that have localized to the spleen[79–81]. CDH16 and CDH17 are expressed in cells typically responsible for water absorption, but their function has yet to be clearly elucidated. In intestinal epithelia,

CDH17 is mainly expressed on the lateral and basolateral side of the cells, but excluded from the adherens junctions and the desmosomes containing CDH1 and the desmosomal cadherins, respectively[74,87]. CDH17's *trans*-interactions across cells are thought to help maintain the width of the interstitial cleft between epithelial cells to regulate water transport[71,72].

Interestingly, CDH17 expression is misregulated in a variety of cancers, with overexpression occurring in gastric, pancreatic, colorectal, and hepatocellular cancer cells[134–142]. CDH17 has also been detected in cells from patients with Barrett's esophagus, a condition where the cells of the esophagus undergo an abnormal change to cellular types usually found in the small and large intestine[85,86]. Given that CDH17 is typically expressed in only a few specialized organs and tissues, this protein can be a therapeutic target and also a useful marker for the variety of cancers in which is expressed. Understanding CDH17's structure and mechanism of adhesion may help design drugs that bind to its extracellular domain to block its function or to deliver other drug components to kill only cancerous cells.

Cadherins are able to mediate adhesion using their extracellular domain, which has tandem extracellular cadherin (EC) repeats[7,8]. The EC repeats have ~110 residues arranged in a seven β-strand Greek-key fold with several conserved sequence motifs of residues involved in binding of three calcium ions at the linker regions between the EC repeats[143]. Calcium binding gives rigidity to the extracellular domain and facilitates adhesion[19–21]. Various subfamily members utilize their EC repeats differently to mediate *trans* adhesion between two cellular membranes. For instance, CDH1 uses a tryptophan on the N-terminal strand of EC1 that is exchanged and inserted into the hydrophobic binding pocket of another EC1[38,42]. This is called the strand-swap mechanism and all classical cadherins use it for their adhesive bond[8]. In addition, there are adhesion mechanisms that involve more EC repeats. Members of the α, β, and γ clustered protocadherins form antiparallel *trans* interactions mediated by EC1-4, and δ1 and δ2 non-clustered protocadherins use the same set of EC repeats, forming a similar antiparallel interaction[22,31,32,51,144]. Cadherins are also capable of forming complexes on the surface of the same cell (*cis* interactions). Classical cadherins use this *cis* interaction to form adhesive patches on the cellular surface[34,42], while clustered protocadherins use a heterophilic mechanism of *cis* interactions to direct neuronal self-avoidance[27,48,133]. All of the antiparallel interactions described above require the most N-terminal repeat, EC1, and do not use a fully overlapped dimer to mediate adhesion.

CDH16 and CDH17 are unique among cadherins as they have seven EC repeats and their EC1-2 and EC3-4 repeats share similarity with one another, as well as with EC1-2 in classical cadherins[74–77,87,93]. Gene structure analyses suggest that CDH16 and CDH17 likely arose from a classical ancestor when the two N-terminal repeats, EC1 and EC2, were duplicated. This is further supported by analyses showing that the EC2-3 linker lacks the sequence motifs with residues necessary to bind calcium ions and might be calcium free altogether[74,76,78]. Additionally, the remainder of the repeats, EC5-7 share sequence similarity with EC3-5 from classical cadherins[74,76].

The short cytoplasmic domains of CDH16 and CDH17 are also unique among cadherins. Classical cadherins have long unstructured cytoplasmic tails (~150 residues) that interact with catenins connected to the cytoskeleton[17,132]. In contrast, the cytoplasmic tails of CDH16 and CDH17 only have ~20-30 residues[74,87], with unclear cytoplasmic binding partners (see Bartolome et al., 2014)[145].

In vitro experiments have revealed that CDH17 forms homophilic *trans-* and *cis-*complexes in a calcium-dependent manner[44,72,87–89]. A CDH17 inhibitory peptide, designed based on expected structural similarities to classical cadherins, interfered with homophilic *trans* adhesion in CACO2 cells forming epithelial layers and resulted in interstitial cleft widening[72]. Experiments also revealed that CDH17 is capable of forming heterophilic *trans* interactions with the classical cadherin CDH1, which is expressed in the same cell types as CDH17[44]. However, a structural basis for how CDH16 and CDH17 interact in *cis* and *trans* to mediate homophilic and heterophilic adhesion is not known.

To better understand the mechanism of adhesion in the 7D family, we determined the x-ray crystal structure of the N-terminal EC1-2 repeats of *Homo sapiens* (*hs*) CDH17, which revealed unique features at its N-terminus relevant for its adhesion mechanism. In parallel, we used bead aggregation assays with the full-length extracellular domain of the protein, a C-terminal truncation series, and an N-terminal truncation of EC1 to determine CDH17's minimal adhesive unit. We found that EC7, but not EC1, is required for bead aggregation.

Our data suggest that CDH17 forms a unique antiparallel homophilic dimer to mediate *trans* adhesion, different from those used by other cadherins.

Results

The N-terminal repeats of CDH17 are structurally unique

Sequence analyses suggest that CDH17 has classical-like cadherin repeats but uses a different mechanism of adhesion than the strand-swap utilized by classical cadherins. Human CDH17 and CDH16 lack the tryptophan residues exchanged by classical EC1 repeats and instead have longer N-termini (~5-8 residues in CDH17 across vertebrates, Fig. 2.1a, Fig. 2.4, and Table 2.2). In addition, EC2 of CDH17 lacks the calcium-binding residues needed to form a canonical EC2-3 linker (Fig. 2.1a and 2.4) and sequences of CDH16 similarly lack these residues. To understand how 7D cadherins structurally differ from classical cadherins in their N-terminal repeats, we produced *hs* CDH17 EC1-2 (p.Q1 to p.A219), crystallized it, and solved its structure refined to 2.15 Å resolution (Table 2.1). This structure has two monomers in the asymmetric unit with chain A encompassing residues p.P11 to p.N215 and chain B encompassing residues p.F5 to p.T213 (residue numbering refers to protein without signal peptide). The associated electron density map (Fig. 2.5) allowed for modeling of all the listed residues except for BC and EF loops in chain A (p.N32 to p.V36 and p.D78 to p.I83 respectively). Since root mean square deviation between the two monomers is low (0.92 Å for Cα atoms), we describe structural features as seen in the most complete monomer (chain B), unless otherwise stated.

The *hs* CDH17 EC1-2 structure revealed both canonical and unique features relevant for function. Repeats EC1 and EC2 fold in typical seven β-strand Greek-key motifs (β-strands labeled A to G; Fig. 2.1b-c), but EC1 has an extra N-terminal β-strand and EC2 has an additional α helix (see below). As expected, the EC repeats are arranged in a linear configuration and there are three calcium ions bound (sites 1, 2, and 3) in a canonical EC1-2 linker region (Fig. 2.1f). This is consistent with the canonical sequence motifs XEX and DRE present in the first repeat (p.17YEG19 and p.63DRE65 in EC1), the DXNDN sequence motif at the linker (p.96DINDN100), and the sequence motifs DXD (p.130DLD132) and XDX (p.192KDM194) present in the second EC repeat (Fig. 2.4). Other loops and secondary structure elements are canonical, highlighting overall similarities with classical cadherin EC repeats.

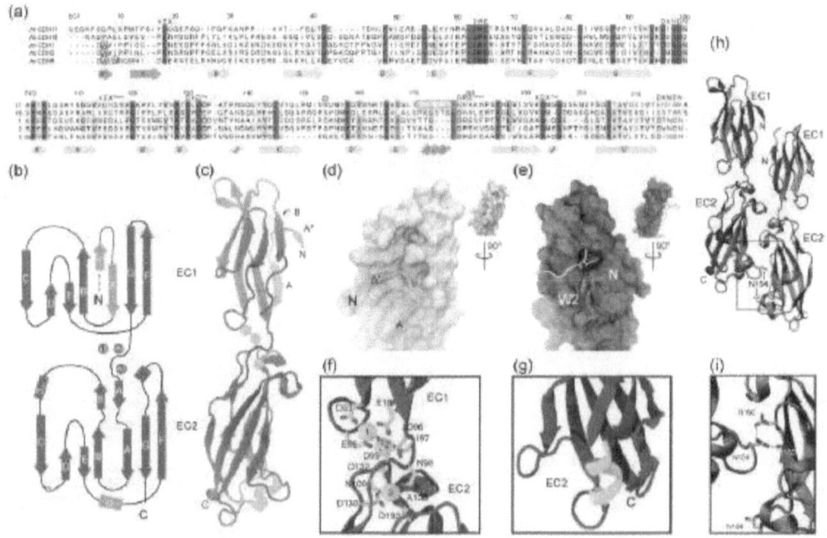

Figure 2.1 Structure of hs CDH7 EC1-2.
(a) Protein sequence alignments of EC1 and EC2 repeats of the 7D cadherins (CDH17, CDH16) and three classical cadherins (CDH1, CDH2, and CDH5) show how these differ at their N-terminus and the EC2 C-terminal tail. Calcium-binding residues are highlighted with their respective sequences on the top. Missing calcium-binding motifs are highlighted in red for CDH17 and CDH16 in EC2. Tryptophan residues that classical cadherins utilize for the strand-swap mechanism are circled in red. The p.N154 residue implicated in disease is denoted by a blue dot[83]. Secondary structure elements of the *hs* CDH17 EC1-2 structure are highlighted below the sequence. (b) Topology of *hs* CDH17 EC1-2. Repeats EC1 and EC2 have canonical Greek-key motifs with atypical features in orange. Residue p.N154 is denoted by a blue dot. (c) Ribbon representation of the *hs* CDH17 EC1-2 structure with calcium ions in green and atypical features in orange. (d) Detail of the *hs* CDH17 EC1 N-terminus showing residues p.F5-p.E18 in ribbon representation over the EC1 surface. (e) Detail of the *hs* CDH1 EC1 N-terminus (PDB: 2O72)[146]. Green strand shows EC1 residues p.D1 to p.E11 extending out from the surface of the rest of EC1, while the silver strand shows the N-terminus of another monomer engaged in the strand-swap mechanism. A tryptophan residue (p.W2) is inserted in the hydrophobic pocket. Insets in panel (d) and (e) show rotated views of EC1 highlighting N-terminal strand location. (f) Detail of the *hs* CDH17 EC1-2 canonical calcium-binding linker. Sidechains of calcium coordinating residues are shown as sticks. Some backbone atoms are omitted for visualization purposes. (g) Detail of the two-turn α-helix in the EC2 E-F loop, highlighted by the orange box in (a). (h) Crystal contacts between two *hs* CDH17 EC1-2 monomers forming a potential *cis* interface as identified by PISA[147]. The interface area is 582.9 Å^2. (i) Detail of the p.N154 residue implicated in disease[83] forming hydrogen bonds with residues p.N162 and p.L160.

Table 2.1 Statistics for Cadherin 17 Structure

Data Collection and Refinement	
Space Group	$P\,2_1\,2_1\,2_1$
Unit cell parameters	
a, b, c (Å)	58.39 81.61 106.81
α, β, γ (°)	90 90 90
Molecules per asymmetric unit	2
Beam Source	APS 24-ID-E
Wavelength (Å)	0.979119
Resolution limit (Å)	50-2.15
Unique Reflections	28521 (1406)
Redundancy	3.6 (3.5)
Completeness (%)	99.5 (99.9)
Average I/σ (I)	10.46 (2.05)
R_{merge}	0.132 (0.730)
CC1/2	0.925 (0.733)
CC*	0.979 (0.92)
Refinement	
Resolution range (Å)	47.53-2.15 (2.19-2.15)
Residues (atoms)	3179
Water Molecules	203
R_{work} (%)	20.6 (26.5)
R_{free} (%)	25.7 (28.2)
Rms deviations	
Bond lengths (Å)	0.0086
Bond angles (°)	1.2686
B factor average	
Protein	32.94
Ligand/ion	24.33
Water	30.89
Ramachandran plot regions	
Most favored (%)	87.5
Additionally allowed (%)	12.5
Generously allowed (%)	0
Disallowed (%)	0
PDB ID code	6ULM

Among the unique, non-canonical features observed in the *hs* CDH17 EC1-2 structure is an additional N-terminal β-strand in EC1 (p.F5 to p.P8), which we labeled A* as it precedes the canonical β-strand A (Fig. 2.1c). The electron density map at the N-terminus of the EC1 monomers is not good enough to model residues p.Q1 to p.K4 in the structure, suggesting some conformational variability at this end. Nevertheless, residue p.F5 and the reminder of the A* β-strand backbone are clearly visible with side chains that were unambiguously modeled (Fig. 2.5a). The A* β-strand runs parallel to and interacts with β-strand B, ending in a turn loop facilitated by two proline residues, p.P8 (highly conserved) and p.P11 (highly conserved among mammals). This double proline A*A turn leads to β-strand A, which runs antiparallel to β-strand B (Fig. 2.1c-d). In contrast to the classical *hs* CDH1 EC1 (Fig. 2.1e and Fig. 2.6c), which uses a short N-terminus that protrudes away from the monomer to engage in a strand-swap mechanism involving β-strand A, CDH17's extended N-terminus lays against and interacts with its own EC1 (Fig. 2.1d-e and Fig. 2.6c). These results suggest that CDH17 EC1 does not engage in classical-like strand swapping interactions.

A second unique feature observed in the *hs* CDH17 EC1-2 structure is a two-turn α helix located in EC2 between β-strands E and F (p.R171 to p.L177). This EC2 EF α helix is facing the EC2-3 linker region and its sequence is well conserved among mammals, with a ~50 residue-long insertion in various fish species (Fig. 2.4, inset). Interestingly, repeat EC2 lacks the XEX, DRE, and DXNDN motifs for a canonical calcium-binding EC2-3 linker region (Fig. 2.1a and 2.4), and the EC2 EF α helix is at the loop that normally would have the DRE motif (Fig. 2.1g), suggesting that the EC2-3 linker region lacks calcium-binding sites 1 and 2. The EC3 sequence has modified DXD (p.244DPG246) and XDX (p.291KDE293) motifs, indicating that it may retain calcium-binding site 3 at the top of EC3. The EC2 EF α helix located at this non-canonical linker region might help rigidify an otherwise flexible, partial calcium-free joint.

Crystallographic contacts reveal possible *cis* and *trans* interfaces

Previous crystallographic structures of cadherin extracellular domains have revealed crystal contacts as possible interfaces for *cis* and *trans* interactions, which in some cases have been validated using *in vitro* biophysical assays as well as *ex-vivo* and *in-vivo* experiments[22,27,28,30–34,38,39,42,43,51,66,144]. Size exclusion chromatography (SEC) experiments with *hs* CDH17 EC1-2 suggest that this fragment is monomeric in solution (Fig. 2.6a-b); yet similar tests have failed to detect weak, physiologically relevant interactions such as those mediating *cis* interfaces in classical cadherins[34]. We used the Proteins, Interfaces, Structures, and Assemblies (PISA) server[147] to identify all possible crystal contacts and evaluate their possible physiological relevance. There are seven crystal contacts with interface areas that range from ~100 to ~583 Å^2, too small compared to an empirical threshold (856 Å^2) that distinguishes biologically relevant interactions[148]. However, physiologically relevant protocadherin interfaces often involve multiples small contacts between various EC repeats (200 to 400 Å^2 per EC repeat)[22,31,149], and the two largest CDH17 EC1-2 interfaces observed in our crystal structure would involve EC repeats beyond EC2. Thus, these two interfaces, discussed below, could become large enough to be physiologically relevant when considering contact contributions from all EC repeats.

The first and largest interface from crystal contacts is formed by two monomers in the asymmetric unit of the crystal arranged in a parallel configuration with an interface area of 582.9 Å^2 (Fig. 2.1h and Fig. 2.4). The orientation of the monomers is such that the β-strands A*, A and B in the EC1 repeats are facing towards each other, and the F and G β-strands face the outside. The monomers are, however, slightly shifted with respect to each other, and the C-terminal part of EC1 of one monomer interacts with the N-terminal part of EC1 of the next monomer, with a similar arrangement observed for the EC2 repeats, and potentially for further EC repeats that would contribute to a large *cis* interface if these were present in the structure. To maintain this arrangement between two parallel membranes, the complex would need to be tilted, as observed for classical cadherins that engage in *cis* interactions in which EC1 interacts in parallel with EC2 (594.8 Å^2)[34]. The possible physiological relevance of the CDH17 *cis* interface is supported by a p.N154S mutation involved in disease[83] located in EC2 at the interface (Fig. 2.1h-i). However, analysis of predicted glycosylation sites suggest that sugars stemming from one of the p.N162 could interfere with this arrangement, as it has been observed for other cadherins[150].

The second largest interface from crystal contacts observed in the *hs* CDH17 EC1-2 structure is formed by an antiparallel overlap of EC2 repeats with an interface area of 565.6 Å^2 (Fig. 2.6d). In this crystal contact, the EC1 repeat would interact with repeat EC3, if it was present, to form an EC1-3 antiparallel interface. The interface area calculated for this crystal contact is for EC2 alone, and it is similar to or larger than the interface area values per repeat observed in other structures of protocadherin interfaces[22]. Thus, an EC1-3 interface would be larger and perhaps physiologically relevant. Again, analysis of predicted glycosylation sites suggest that sugars stemming from p.N127 could interfere with the EC2-EC2 contacts (Fig. 2.4 and 2.6d). Other crystal contacts detected by PISA had small interface areas and were incompatible with simple *cis* and *trans* interactions given the arrangement of the EC repeats (Fig. 2.6e-i).

Overall, sequence analyses and our *hs* CDH17 EC1-2 structure indicate that CDH17 does not use an EC1 strand-swap mechanism to mediate *trans* adhesion and suggest plausible *cis* and *trans* interactions that we further tested using bead aggregation assays.

CDH17 requires repeat EC7, but not EC1, for *trans* adhesion

Bead aggregation assays carried out with full-length and truncations of ectodomains have been shown to be a robust semi-quantitative method useful in determining the minimum binding unit of *trans* adhesion for cadherins[22,31,45,67]. Although previous studies have characterized the homophilic *trans* interaction mediated by CDH17, all prior experiments used full-length extracellular fragments. Based on the crystallographic interfaces observed in the crystal structure of *hs* CDH17 EC1-2, we designed a library of truncations that eliminated EC repeats from the C-terminal end of the extracellular domain sequentially, ranging from the full-length extracellular domain down to EC1-2 only. These constructs were expressed in HEK293T using the native signal peptide and a C-terminal Fc-tag.

Results from bead aggregation assays using the entire extracellular domain of *hs* CDH17 (EC1-7 Fc) confirmed its ability to mediate *trans* adhesion in the presence of calcium (Fig. 2.2a). Small bead aggregates can be observed immediately after mixing (0 min) with their size increasing after 60 min, and with larger clumps forming after rocking for 1 or 2 min (Fig. 2.7a). Aggregation was abolished in the presence of a calcium chelator (2 mM EDTA) (Fig. 2.2b and 2.7h) indicating that it is calcium dependent as observed for other cadherins. To identify the CDH17 minimal adhesive unit, aggregation assays were then performed with CDH17 EC1-6 Fc (Fig. 2.2c and Fig. 2.7b) and other truncations down to EC1-2 Fc (Fig 2.7c-f). Lack of aggregation was observed upon eliminating EC7 from the construct, as in EC1-6 Fc, even after rocking (Fig. 2.2c and 2.7b), with results similar to the EDTA control (Fig. 2.2d and Fig. 2.7i). All further C-terminal truncations showed similar results to EC1-6 Fc in the presence of calcium. Previous binding assays performed using other cadherins typically show a similar shift from aggregation to no aggregation when a single essential EC repeat is removed[22,31]. Our results indicate that CDH17 requires EC7 at the C-terminus to mediate *trans* adhesion. We then proceeded to determine if EC1 at the N-terminus was necessary for adhesion to occur.

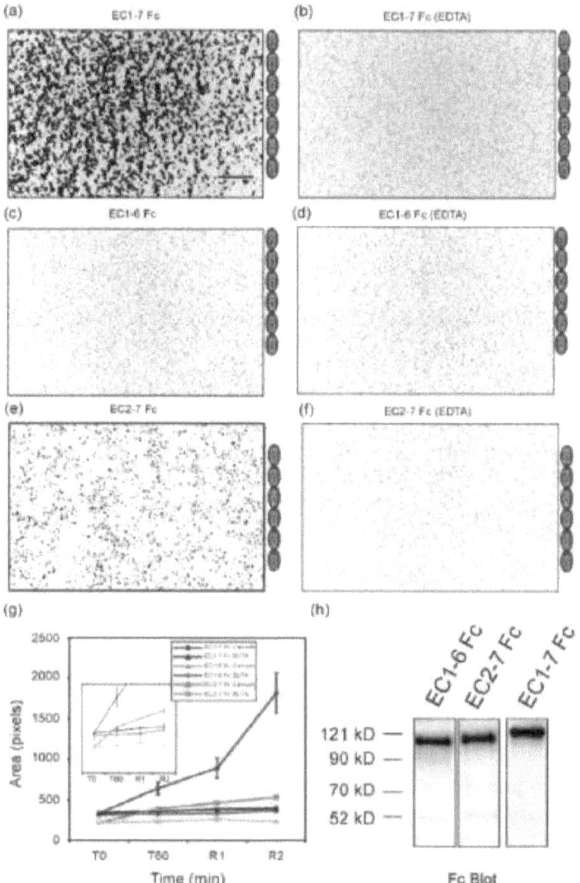

Figure 2.2 Bead aggregation assays for CDH17.
(a-f) Protein G beads coated with the full-length CDH17 EC1-7 Fc ectodomain (a-b), C-terminal truncation EC1-6 Fc (c-d), and N-terminal truncation EC2-7 Fc (e-f). Each image shows beads observed after 60 min followed by rocking for 2 min in the presence of 2 mM CaCl$_2$ (a, c, e) or 2 mM EDTA (b, d, f). Bar – 500 µm. Images in (a)-(f) are also shown in Fig. 2.7 (R2). (g) Aggregate size for the full-length CDH17 EC1-7 Fc ectodomain, the truncated EC1-6 Fc, and EC2-7 Fc at the start of the experiment (T0), after 60 min (T60) followed by rocking for 1 min (R1) and 2 min (R2). Inset shows zoom in of aggregate area up to 750 pixels. Error bars are standard error of the mean ($n = 9$ for full-length CDH17 EC1-7 Fc, $n = 5$ for EC1-6 Fc, and $n = 2$ for EC2-7 Fc). (g) Western blot shows expression and secretion of the corresponding fragments.

Elimination of the most N-terminal repeat (EC1) is typically effective in demonstrating the necessity of the N-terminus of cadherins in the formation of the *trans* interaction both in cells[67] as well in bead aggregation assays[22,31]. Thus, we prepared a construct that eliminated EC1 from between the signal peptide and EC2 (EC2-7 Fc) and performed bead aggregation assays (Fig. 2.2e-f and Fig. 2.7g, j). In the presence of calcium, the EC2-7 Fc fragment still facilitated the formation of bead aggregates (Fig. 2.2e). These results suggest that EC1 is not essential for CDH17-mediated *trans* adhesion.

Analysis of SEC experiments using hexa-histidine tagged protein fragments expressed using Expi293F cells (*hs* CDH17 EC1-7, EC1-6, and EC2-7) revealed that these are stable and well-folded (Fig. 2.7n). Additionally, only properly expressed, secreted, and processed protein with a natively folded Fc tag should be able to interact with the protein G beads in aggregation assays, meaning that changes in bead aggregation should not be caused by misfolding or instability when removing EC repeats. Therefore, our bead aggregation assays suggest that EC7, but not EC1, is essential for CDH17-mediated *trans* adhesion.

CDH17 mutations have mixed effects on bead aggregation
Disease-related mutations in the extracellular domains of cadherins often abolish adhesive function, either directly or by impairing calcium binding or proper folding[22,24,33,66]. Thus, bead aggregation assays with protein fragments carrying missense mutations involved in disease can be used to understand *trans* and *cis* adhesive mechanisms. There is only one known missense mutation in CDH17 implicated in disease, p.N154S[83], which might interfere with the *cis*- rather than the *trans*- interaction mediated by CDH17. Beads still

aggregated when assays were carried out with the mutated ectodomain of CDH17, EC1-7 Fc N154S (Fig. 2.8a), suggesting that this mutation causes effects unrelated to *trans* adhesion. In addition, if the p.N154S mutation were to alter *cis* dimerization, our results would indicate that *trans* adhesion does not require *cis* interactions.

One additional mutation site was selected based on its possible involvement in CDH17 *trans* adhesion. A tryptophan residue in EC3, p.W217, equivalent to the EC1 p.W2 residue involved in the tryptophan-mediated strand-swap of classical cadherins, was mutated to an alanine (W217A) and to an arginine (W217R). The CDH17 EC1-7 Fc W217A and W217R protein fragments did not mediate bead aggregation in the presence of 2 mM calcium (Fig. 2.8b-c), as seen for EDTA controls. These mutations might destabilize the ectodomain or might directly impair *trans* adhesion, thus implicating EC3 in binding interactions.

Sequence analysis highlight diversity across species and family members
To better understand what domains and specific residues are relevant for adhesive function, we performed sequence alignments and determined CDH17 conservation across over 200 species, ranging from mammals to birds, reptiles, fish, and amphibians. We also compared *hs* CDH17 and CDH16 sequences to establish whether adhesive mechanisms are shared across the 7D-cadherins.

An analysis of conservation across species reveal that CDH17 EC1-2 is not well conserved when considering all vertebrate species (Fig. 2.3a). Mapping of conservation on the structure of EC1-2 reveals that the most conserved sites (41 residues) are calcium-binding residues (7 residues) or those forming the hydrophobic core (29 residues) (Fig. 2.9a,e,m). The remaining 5 highly conserved residues face the outside of the protein (2.3% of all residues in the structure). In contrast most of the lowest conserved residues (28 out of 30) were surface exposed (13.3% of all residues). Given the unusual insertion in fish EC2 sequences (Fig. 2.4 and 2.10), we focused sequence analysis on the mammalian CDH17 proteins. We found that CDH17 EC1-2 is overall more conserved among mammals (Fig. 2.3a). In addition to residues forming the calcium-binding sites (10 residues) and hydrophobic core (37 residues), residues on the surface of CDH17 EC1-2 were found to be more highly conserved (12 residues, 5.7%) when only considering mammalian species (Fig. 2.9b,f,m). Conservation of residues at the possible EC1-2 *cis* interface, including the likely glycosylated p.N162, is generally higher when considering only mammalian species (Fig. 2.9i-l).

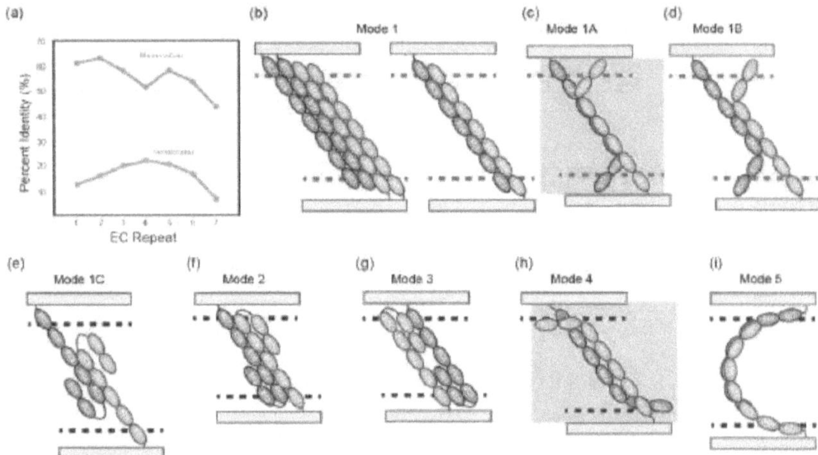

Figure 2.3 Sequence conservation of CDH17 EC repeats and possible adhesive interfaces.
(a) Percent identity of individual repeats from Fig. S1 sequence alignment for mammalian species in blue, and for vertebrates including mammals, reptiles, birds, and fish, in orange. (b-g) Schematics of possible modes of *trans* interaction for CDH17, including a full overlap of the EC1-7 (Mode 1 with p.N154 highlighted in blue), as well as modes that involve a slightly flexible EC2-3 linker (1A with interactions between EC2 and EC6 and 1B in which EC2 is dispensable), or a fully flexible EC2-3 linker (Modes 1C, 2, and 3). (h) A modification of Mode 1A where EC2 and EC7 interact and EC1 is dispensable (Mode 4). (i) Mode 5 involving an EC1-5 overlap, with contacts between EC2 and EC4 repeats[151]. Dashed lines represent the location of the membrane if EC7 is truncated. Green background highlights modes 1A and 4, which are the most likely interfaces given the results presented here and in Yui et. al.[151].

We also performed conservation analysis on CDH17 EC6-7 sequences, as bead aggregation assays suggest that EC7 is essential for adhesive function. Results were similar to those obtained for EC1-2 with higher conservation in mammals than when fish, reptiles, and amphibians sequences are included. Conservation was mapped to a homology model of EC6-7 generated from repeats EC4-5 from desmocollin 1 (5IRY)[30] (Fig. 2.9c,d,g,h). Three residues are highly conserved on the surface of the EC6-7 homology model when considering all species, with an additional seven residues being surface exposed when only mammal sequences are used in the analysis (Fig. 2.9m).

An analysis of sequence conservation per EC repeat for EC3, EC4, and EC5 shows a similar trend with high percent identity for mammalian species and low percent identity across all vertebrate sequences analyzed (Fig. 2.3a and 2.4). Interestingly, percent identity for EC1-2 is highest among the repeats in mammals but not in vertebrates, while EC7 has the lowest percent identity regardless of which species are included. The variation in percent identity across the EC repeats (from 43% to 63%) is less dramatic than what is observed for longer cadherins like CDH23, where the middle EC repeats are typically less conserved[24]. This suggests that the middle repeats of CDH17 might be the most functionally relevant.

Adhesive mechanisms are typically the same within cadherin subfamilies. CDH16 is another 7D-cadherin and has sequence features similar to those of CDH17, namely, an elongated EC1 N-terminus without the tryptophan involved in classical cadherin adhesion, an atypical EC2-3 linker, seven EC repeats with EC7 featuring a classical-like set of disulfide bonds, and a short cytoplasmic tail. As expected, a comparison of EC repeats within the CDH17 and CDH16 proteins reveals that there is high sequence identity between EC1 and EC3, and EC2 and EC4 (Fig. 2.11a,b). However, CDH16 lacks canonical calcium-binding linker regions in EC5-6 and EC6-7 (Fig. 2.11c), which could render its extracellular domain flexible. It is possible that CDH16 uses a different adhesive mechanism. Overall, sequence comparisons across species and family members suggest involvement of middle repeats in function and diversity in modes of adhesion for the 7D-cadherins.

Discussion

Previous work on CDH17 focused on understanding its adhesive functions at the cellular level, with some insights about underlying molecular mechanisms stemming from comparisons to classical cadherins and from single-molecule experiments testing homophilic interactions with itself and heterophilic interactions with CDH1[44,72,89,152]. Our structure of EC1-2 demonstrates that CDH17 is incapable of forming the EC1 tryptophan-mediated strand-swap, the mechanism used by all classical cadherins, as its EC1 N-terminus is extended and forms the A* strand. A tryptophan residue in EC3 could mediate a classical-like *trans* interaction[44], but a single-point contact at EC3-EC3 would be incompatible with the results of our bead aggregation assays indicating that EC7 is required for *trans* adhesion. Similarly, an anti-parallel EC1-3 interface suggested by crystal contacts in our structure is incompatible with results from our bead aggregation assays and with predicted glycosylation. Thus, a different mechanism potentially involving overlap of the entire CDH17 ectodomain might underlie adhesion.

Our data does not provide details of the arrangement that the CDH17 extracellular domain adopts to mediate adhesion. However, there are several modes of adhesion that are compatible with our results. In the first mode, the extracellular domains of CDH17 protruding from one cell would arrange in a parallel linear conformation compatible with the *cis* interaction observed in our crystal structure that might be regulated by glycosylation at p.N162 and perhaps physiologically relevant as suggested by the effects of the p.N154S disease mutation [83]. This parallel dimer would form an antiparallel interaction with the CDH17 ectodomain from an opposing cell (Fig. 3b). This type of adhesive mode would be similar to what has been observed for classical cadherins (CDH1) where *cis* and *trans* interactions form Velcro-like patches mediated by tip-to-tip interactions of EC1 repeats [34]. However, CDH17 would use more EC repeats and would form zipper-like sections since the *cis* dimer does not allow for propagation of the interaction in a lattice-like fashion. This arrangement would be similar to what is observed for the extracellular domains of clustered protocadherins forming neuronal contacts[27,48], but with full overlap of the CDH17 ectodomain both in *cis* and *trans* (Fig. 2.3b).

If a full overlap of CDH17 ectodomains was mediating adhesion (Mode 1) and EC1 interacted with EC7, EC2 with EC6, EC3 with EC5, and EC4 with EC4, respectively, we would have expected lack of bead aggregation when using the EC2-7 Fc protein fragment.

However, elimination of EC1 did not affect bead aggregation, suggesting that at least two alternative binding modes, Modes 1A and 1B (Fig. 2.3c,d), could account for our experimental results. In these models, internal repeats (EC2-6 and EC3-5) are involved in interactions essential for binding, but EC1 is not necessary for adhesion and elimination of EC7 results in steric hindrance with the Fc tag and protein G beads, thus preventing complex formation and bead aggregation. Modes 1A and 1B involve EC2-6 and EC3-5 as the minimal repeats necessary for the *trans*-interaction to occur. Our data currently suggest that either of these modes is feasible.

Similar to Mode 1B is Mode 1C (Fig. 2.3e), where flexibility at the EC2-3 linker would allow repeats EC1-2 to fold back without participating in the *trans* interaction and without causing steric hindrance when repeat EC7 is eliminated. This mode conflicts with our bead aggregation assays, which suggest EC7 is necessary for the interaction. Alternatively, the minimal overlapping unit might be EC3-7 (Mode 2; Fig 2.3f). If the EC2-3 linker were flexible such that EC1-2 is folded back and not clashing with the cell membrane (or the Fc tag and protein G beads in our binding assays), then interactions of EC7 with EC3, EC6 with EC4, and EC5 with EC5 would mediate *trans* adhesion.

Our data indicating that CDH17 EC2-7 Fc mediates bead aggregation does not rule out involvement of EC2 in the *trans* interaction. If EC1-2 is flexible relative to the EC2-3 linker, an interaction like that of Mode 3 (Fig. 2.3g) might be possible where EC2 interacts with EC7 and EC1 interacts with EC6. This mode might have weak interactions between EC1 and EC6, such that deletion of EC1 is inconsequential.

Several modes of CDH17 adhesion proposed here involve a flexible EC2-3 linker. Non-canonical linker regions that are calcium-free or partially calcium-free have been reported. Notably, in PCDH15, linker regions EC2-3, EC3-4, and EC5-6 are partially calcium-free and exhibit some flexibility[25,153]. The PCDH15 EC9-10 linker is completely calcium-free and is bent, as determined by structures and small-angle X-ray scattering[23]. Similarly, the *Drosophila melanogaster* N-cadherin (*Dm* CDH2) EC2-3 linker is calcium-free and bent[26], and a structure of a calcium-free T-cadherin EC1-2 fragment shows a completely folded, U-shaped, conformation for the repeats[28]. Although previous experiments have demonstrated that calcium binding is essential for providing rigidity to the cadherin EC linker, the bent conformations observed are stable. The CDH17 EC2-3 linker might be calcium-free or might have a calcium ion bound in site 3, allowing for EC1-2 to be flexible and kinked (Modes 1A and 1B) or to fold back to, in some cases, interact with other EC repeats (Modes 1C, 2, 3, and 4; Fig. 2.3). Sequence conservation analysis does not provide information that could favor one mode over the others.

Recent work presented a structure of CDH17 EC1-4 forming an anti-parallel complex where the EC2 repeat of one monomer interacts with EC4 from the other, and EC3 is not directly involved in the binding interface[151]. Based on the overlap of the EC repeats in the crystal structure, an EC1-5 *trans* interaction is suggested (Mode 5). Intriguingly, the EC1-

4 protein fragment was mostly monomeric in their SEC multi-angle light scattering experiments at low protein concentration and dimeric in analytical ultracentrifugation experiments at higher concentrations. The F224A mutation at the proposed interface abolished dimerization and cell aggregation, but also compromised protein stability. Importantly, the EC1-5 and EC3-7 protein fragments did not facilitate cell aggregation[151]. These last results are consistent with our bead aggregation assays indicating that repeat EC7 is necessary for adhesion and strongly favoring Modes 1A, 2, 3 and 4. The CDH17 EC1-4 structure also confirms the predicted calcium-free EC2-3 linker, with molecular dynamics simulations predicting some flexibility at this linker[151], suggesting that Modes 1A and 4 are most consistent with experimental and simulation data. We cannot rule out, however, that removal of EC7 causes clashes with the membrane[151] or allosterically prevents *trans* interactions without the need of having a direct interaction of EC7 with another EC repeat (Mode 5). Its low sequence conservation is consistent with a "spacer" role.

The most likely adhesive modes described above involve antiparallel arrangements in which EC2, EC3, and EC7 may contribute to the *trans* interaction. In vertebrates, the largest overlap of cadherin ectodomains mediating *trans* adhesion involves repeats EC1 to EC4 (clustered, δ1-, δ2-, Fat and Dacshous protocadherins)[22,29,31,32,144,154]. However, larger overlaps have been observed in invertebrate cadherins. The *Dm* CDH2 is a homologue of the vertebrate N-cadherin (CDH2) and has 16 EC repeats in its extracellular domain[26]. Analytical ultracentrifugation experiments suggest that EC1 to EC9 are required for the *trans* interaction of *Dm* CDH2 to occur. Structures of its *trans* interaction have yet to be elucidated, but *Dm* CDH2 performs the same function as CDH2 in vertebrates. A structure of *Dm* CDH2's N-terminal repeats EC1-EC4 revealed that this protein does not mediate adhesion through the tryptophan strand-swap like vertebrate classical cadherins[26]. Interestingly, it is likely that cadherins with long ectodomains and multiple non-canonical linker regions, like *Dm* CDH2, FATs, DCSH, and other invertebrate cadherins might adopt a globular conformation reminiscent of what is observed for the Dscams[46,155]. CDH17 only has one non-canonical binding linker and thus is likely to mediate adhesion using the modes discussed above, with specific faces of EC2 and EC7 mediating the *trans* interaction.

Methods

Cloning of CDH17 constructs

DNA encoding for *hs* CDH17 (Harvard PlasmidID: HsCD00419124) with two natural variations (p.K93E and p.E712D) was used as a template in all polymerase chain reaction amplification experiments. The sequence encoding for EC1-2 was sub cloned into *NdeI* and *XhoI* restriction sites of the pET21a vector, which was used for bacterial expression of protein used in crystallization experiments. In addition, DNA constructs for the full-length ectodomain of CDH17 and a series of C-terminal truncations of it were created for mammalian expression of protein used in bead aggregation assays. Truncation locations were chosen based on the DXNDN linker sequence or similar ends for each EC repeat (Fig. 2.4). The native signal sequence N-terminal to EC1 was kept in all constructs used for mammalian expression. The full-length ectodomain and truncations were cloned into a modified CMV vector that adds a C-terminal Fc-tag or a hexa-histidine tag (Jontes laboratory, OSU) using *XhoI* and *KpnI* restriction sites. Mutations were generated using the QuikChange Lightning kit (Agilent). All DNA constructs were sequence verified.

Bacterial expression and purification of CDH17 fragments

Rosetta 2 (DE3) cells (Novagen) were transformed with the *hs* CDH17 EC1-2 construct, cultured in terrific broth (TB), and induced at $OD_{600} \sim 0.45$ with 1 mM isopropyl β-d-1-thiogalactopyranoside (IPTG) at 30°C overnight. The cells were then lysed by sonication using denaturing buffer (20 mM TrisHCl [pH 7.5], 6 M guanidine hydrochloride [GuHCl], 10 mM $CaCl_2$ and 20 mM imidazole) and centrifuged. The resulting cleared lysates were incubated with Ni-Sepharose beads (GE Healthcare), washed twice with denaturing buffer, and eluted with the denaturing buffer supplemented with 500 mM imidazole. The purified *hs* CDH17 EC1-2 protein was refolded overnight at 4°C using molecular weight cut-off (MWCO) 2000 membranes in a dialysis buffer containing 20 mM TrisHCl (pH 8.0), 5 mM $CaCl_2$, 150 mM KCl, 50 mM NaCl, and 400 mM arginine. The refolded protein was further purified on a Superdex 200 16/600 column (GE Healthcare) in 20 mM TrisHCl (pH 8.0), 2 mM $CaCl_2$, 150 mM KCl, and 50 mM NaCl. The protein was concentrated to ~ 8 mg/mL for crystallization trials (Vivaspin 10 kDa).

Crystallization, data collection, and structure determination of *hs* CDH17 EC1-2

Crystals were grown via vapor diffusion using the sitting drop method at 4°C. An initial dataset was obtained from a crystal grown in 0.1 M Na Acetate pH 5, 0.06 M $CaCl_2$, and

25% 2-methyl-2,4-pentadiol (MPD). The crystal was fished without additional cryo-buffer and cryo-cooled in liquid nitrogen. A diffraction dataset was obtained at the Ohio State University Campus Chemical Instrument Center using a MicroMax-003 x-ray generator and a PILATUS R200K Hybrid Pixel Array Detector. Indexing and scaling were done in HKL3000[156] in the space group $P2_1$. This initial dataset was solved using Mr. BUMP[157] in CCP4[158] using a model of E-cadherin EC1-5[34] (PDB: 3Q2V). This model was subsequently refined at 2.85 Å resolution using REFMAC5[159]. A second, higher resolution dataset was obtained from a crystal grown in 0.1 M HEPES (pH 7), 0.1 M KCl, and 15% PEG 5000 monomethyl ether (MME). This crystal was fished and cryo-cooled in a crystallization buffer with 25% glycerol. The dataset was obtained at the advanced photon source (APS NECAT 24-ID-E) and was indexed and scaled in HKL2000[160] using the $P2_12_12_1$ space group. The new structure, refined at 2.15 Å resolution in REFMAC5[159], was solved using Phaser [161] with the initial model of *hs* CDH17 EC1-2 from the lower resolution dataset. The structure contains two molecules of *hs* CDH17 EC1-2 in the asymmetric unit, with chain B having the most complete modeling of residues. The structure for *hs* CDH17 EC1-2 has been deposited into the Protein Data Bank (PDB) with code 6ULM. Data and refinement statistics are in Table 1.

Binding Assays

Bead aggregation assays were performed as previously described[22,67,162]. Briefly, HEK293T cells were transfected with the CDH17-Fc fusion constructs via calcium-phosphate transfection[163–165]. To this end, a solution of 10 µg of DNA and 250 mM $CaCl_2$

was prepared and added dropwise to 2X HEPES buffered saline (HBS) while mildly vortexing. This solution was then immediately added drop-wise to a 100 mm dish of HEK293T cells (80% confluency) cultured in Dulbecco's Modified Eagle Medium (DMEM) with fetal bovine serum (FBS), L-glutamine (L-glut), and penicillin-streptomycin (PenStrep). For each construct, two plates of cells were used. After 20 h post-transfection, cells were rinsed twice in 1X phosphate buffered saline (PBS) and allowed to grow in serum-free DMEM with L-glut and PenStrep for two days prior to media collection. The media containing the secreted Fc fusion proteins was collected and concentrated to a volume of 500 µL (Amicon 10 kDa concentrators). This was mixed with 1.5 µL of protein G Dynabeads (Invitrogen) and incubated with rotation at 4°C for 2 h. Beads were washed with binding buffer (50 mM Tris HCl [pH 7.5], 100 mM NaCl, 10 mM KCl, and 0.2% bovine serum albumin [BSA]) and split between two tubes, one containing 2 mM $CaCl_2$ and the other 2 mM EDTA. Beads were allowed to aggregate on glass depression slides in a humidified environment and images were captured using a Nikon Eclipse Ti microscope with a 10X objective at 0 and 60 min, and after rocking the slides for 1 and 2 min at 8 oscillations/min. The area of the aggregates was quantified in pixels using ImageJ[166]. Plots show average aggregate size ± standard error. Number of independent biological replicates were nine for EC1-7 Fc, five for all C-terminal truncations (EC1-6 Fc, EC1-5 Fc, EC1-4 Fc, EC1-3 Fc, and EC1-2 Fc), two for EC2-7 Fc and three for all EC1-7 Fc mutants (N154S, W217A, and W217R).

Western Blots

Western blots were performed for Fc-tagged proteins to confirm expression and secretion of protein fragments. Samples were mixed with 4X SDS with β-mercaptoethanol, boiled for 5 min, and loaded onto gels for SDS-PAGE (BioRad) experiments. Proteins were electroblotted onto PVDF membrane (GE Healthcare), blocked with 5% nonfat milk in Tris-buffered saline (TBS) with 0.1% Tween-20 for 1 h, and incubated overnight at 4°C with goat anti-human IgG (1:200; Jackson ImmunoResearch Laboratories, Catalog-109-025-003, Lots 117103, 135223). After washing with 1X TBS, the membranes were incubated with mouse anti-goat HRP-conjugated secondary antibody (1:5000; Santa Cruz Biotechnology, Catalog-sc-2354, Lot-A2017) for 1 h at room temperature. Membranes were washed in 1X TBS and developed using the ECL Select Western Blot Detection kit (GE Healthcare) for chemiluminescent detection (Omega Lum G). Western blots were performed using samples collected immediately after media collection and concentration as well as using protein samples collected from the Protein G beads following aggregation assays. The iBright Prestained ladder (ThermoFisher Scientific) was used for each Western blot. This ladder does not react with the primary or secondary antibody. Immediately following the capture of the chemiluminescent image, another image with white light (Fig. S4l) was taken on the same machine (Omega Lum G) with the same PVDF membrane in order to have images of the Western blot and ladder in the same position for validation purposes.

Mammalian expression of CDH17 EC1-7, EC1-6, and EC2-7

The CDH17 EC1-7, EC1-6, and EC2-7 His fragments were expressed in Expi293F™ cells (ThermoFisher Scientific). Transfection was done with 30 µg of plasmid and the ExpiFectamine™ 293 transfection kit. Prior to addition of the DNA mixture, 5 µM kifunensine was added to prevent the formation of elongated glycosylation sites. Cells were incubated at 37°C in 8% CO_2 at 125 RPM. Three days post-transfection, the cells and media were centrifuged to pellet the cells and collect the supernatant to which 10 mM $CaCl_2$ was added. The supernatant was dialyzed overnight in a buffer containing 20 mM TrisHCl (pH 8.0), 150 mM KCl, 50 mM NaCl, and 10 mM $CaCl_2$ using MWCO 2000 membranes to remove the EDTA present in the media. The fragments were then purified by nickel affinity chromatography in native buffer conditions (20 mM TrisHCl [pH 8.0], 300 mM NaCl, 10 mM $CaCl_2$, and 20 mM imidazole). The protein was eluted with the native buffer supplemented with 500 mM imidazole. The eluted protein was further purified by SEC on a Superose 6 10/300 (GE) column in buffer containing 20 mM TrisHCl (pH 8.0), 150 mM KCl, 50 mM NaCl, and 2 mM $CaCl_2$.

Sequence Alignments and model generation.

Alignments of the EC1-2 repeats of *hs* CDH17 (NP_004054.3), CDH16 (NP_004053.1), CDH1 (NP_004351.1), CDH2 (NP_001783.2), and CDH5 (NP_001786.2) were performed in MUSCLE[167]. The alignments were arranged in JalView[168], and colored using the % identity for sequence identity with a conservation threshold cutoff of 40% for all alignments unless otherwise noted. The numbering on these alignments corresponds to the residue numbering in the human construct without signal peptide. Signal peptide cleavage sites were determined using the SignalP-5.0 server[169]. Protein sequence alignments of individual CDH17 EC repeats were performed as above from 20 different species from NCBI (Table 2.2). Additional sequences were collected from NCBI (Table 2.3) and alignments of EC1-2 and EC6-7 were generated in MUSCLE [167] before running them through the Consurf[170–174] server with the EC1-2 structure and the EC6-7 model described below. The corresponding conservation values per residue (1 = lowest, 9 = highest) were used to generate a heat map of conservation in the structural models. The Swiss-PDB viewer [175] was used to determine surface residues with a cutoff of 20% solvent accessible surface area to generate a list of residues on the surface, forming the hydrophobic core, or involved in calcium binding. Protein sequence alignments of 37 fish species (Fig. 2.10) were generated for the insertion seen in EC2 using MUSCLE[167] and JalView[168], as described above. Sequence alignments of individual EC repeats of *hs* CDH17 and CDH16 were performed in MUSCLE[167] and percent identity between repeats was calculated in Geneious Prime 2020.1.2 to generate a heat map of sequence identity. A model of *hs* CDH17 EC6-7 was generated using (PS)2 v.3[176], the CDH17 protein sequence from p.D537 in EC5 to p.C750 in EC7, and the desmocollin 1 structure (PDB: 5IRY)[30] as a template.

Figure generation software

All protein structures were rendered in Visual Molecular Dynamics (VMD) version 1.9.3[177] with the exception of the electron density maps, which were produced using the CCP4 program FFT and PyMol version 2.4.0.

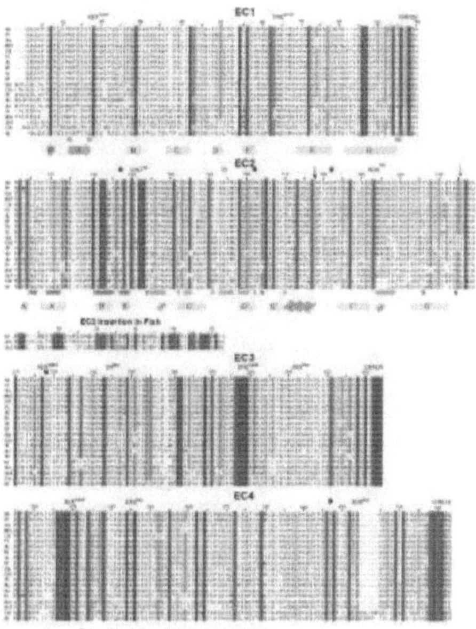

Figure 2.4 Sequence alignments of individual CDH17 EC repeats.
Multiple sequence alignments comparing each EC repeat of CDH17 from 20 different species (Table 2.2). Each alignment is colored by percent identity, with white being the lowest percent identity and dark blue being the highest. Sites of N-linked glycosylation are denoted by a dark blue circle and sites of the natural variants p.K93E and p.E712D are denoted by green boxes in the human sequence. Site of the p.N154 residue implicated in disease is marked with a blue circle. Secondary structure elements observed in the crystal structures of *hs* CDH17 EC1-2 are illustrated below the respective repeats. Calcium-binding motifs are indicated above the sequences, which are numbered according to the human protein without the signal peptide. Residues with ≥ 20% buried surface area in the EC1-2 *cis* interface (Fig. 2.1h) are denoted by orange bars and the same for the EC2-EC2 *trans* interface in blue (Fig. 2.6d. An arrow in EC2 indicates the location of a ~ 50 amino acid insertion observed in all fish species, which has been aligned for included species below EC2. A sequence alignment of this insertion for a variety of fish is in Fig. 2.10. An arrow in EC7 indicates the putative end of this repeat based on structures of classical cadherins[30,34]. Bead aggregation experiments were performed with the entire extracellular domain up to p.V763. Truncation boundaries for each EC repeat were chosen according to the alignments shown in this figure. An arrowhead in blue in EC2 denotes the beginning of EC3 based on the *hs* CDH17 structure. Species are abbreviated as follows: *Homo sapiens (hs), Pan troglodytes (Pt), Rattus norvegicus (Rn), Mus musculus (mm), Canis lupus familiaris (Clf), Felis catus (Fc), Acinoyx jubatus (Aj), Bos Taurus (Bt), Sus scrofa (Ss), Delphinapterus leucas (Dl), Gallus gallus (Gg), Corvus cornix cornix (Ccc), Aptenodytes forsteri (Af), Anolis carolinensis (Ac), Alligator sinensis (As), Danio rerio (Dr), Salmo salar (Ss2), Astyanax mexicanus (Am), Callorhinchus milli (Cm),* and *Xenopus tropicalis (Xt).* Species were chosen based on sequence availability and taxonomical diversity. Accession numbers and species can be found in Table 2.2.

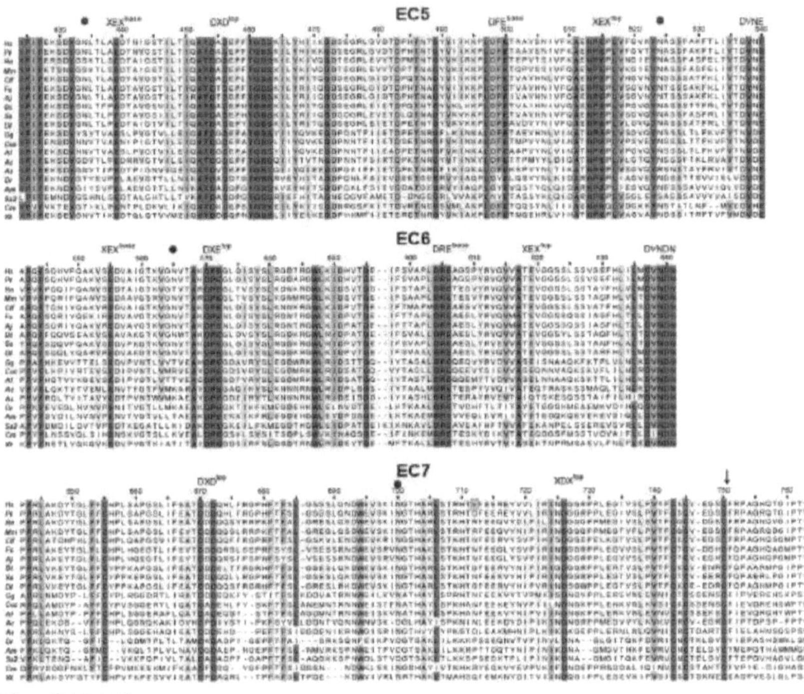

Figure 2.4 (cont.)

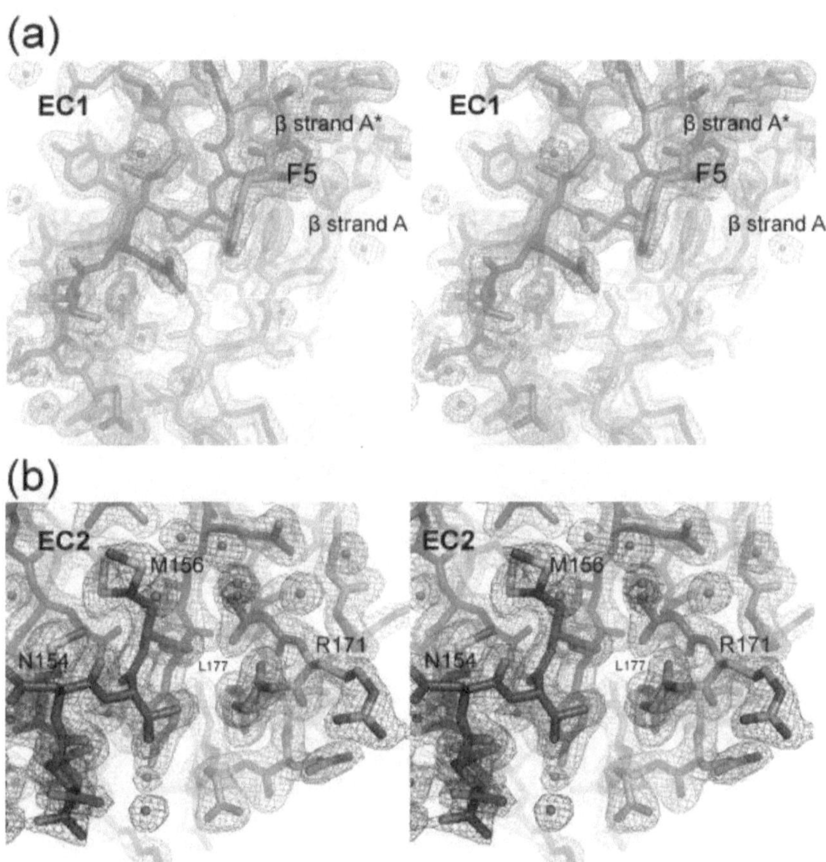

Figure 2.5 Electron density for unique elements of hs CDH17 EC1-2.
(a) Stereo view of EC1 detail showing β-strands A* and A in orange, folded back against EC1. (b) Stereo view for the C-terminal side of EC2 with a two-turn α-helix, shown in orange. Residue p.N154 implicated in disease and a double conformer for residue p.M156 are highlighted. Electron density is shown at 1 σ with a carve of 1.6. Red spheres indicate water molecules present in the electron density.

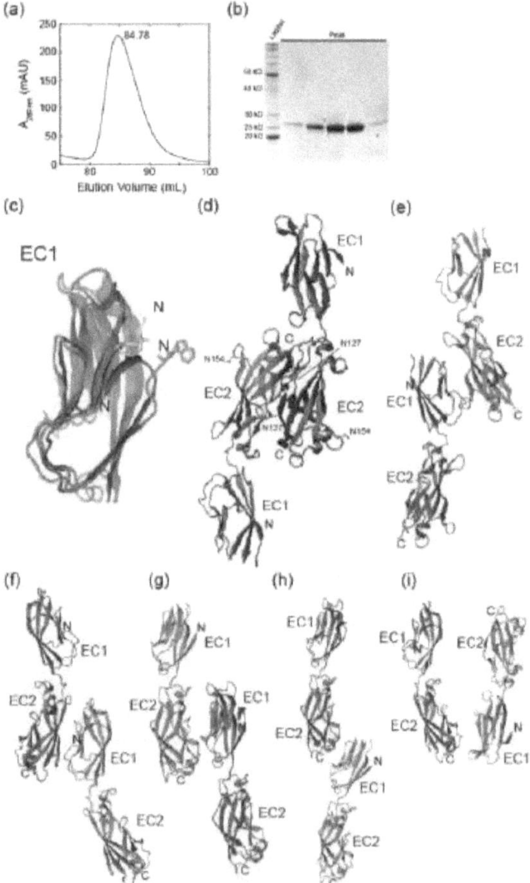

Figure 2.6 Purification and crystal contacts for hs CDH17 EC1-2.
(a) Elution profile of a SEC experiment for *hs* CDH17 EC1-2 showing a monomeric peak at elution volume 84.78 mL using a Superdex S200 column. (b) SDS-PAGE analysis of monomeric peak of *hs* CDH17 EC1-2 from SEC experiment. Lane 1 shows molecular weight standards. Lanes 2-6 show fractions from the monomeric peak. Sample intensity corresponds to peak intensity as equal volumes for each fraction were run on the gel. (c) Detail of the p.N154 residue implicated in disease [83] forming hydrogen bonds with residues p.N162 and p.L160. (d-j) Crystal contacts between two monomers of *hs* CDH17 EC1-2 as identified by PISA[147]. Residue p.N154, implicated in disease, is highlighted in (d) and (e). A predicted N-glycosylation site, p.N127 is highlighted in (e). Interface areas are 582.9 Å^2 (d), 565.6 Å^2 (e), 464.2 Å^2 (f), 454.0 Å^2 (g), 452.0 Å^2 (h), 183.4 Å^2 (i), and 100.0 Å^2 (j) respectively.

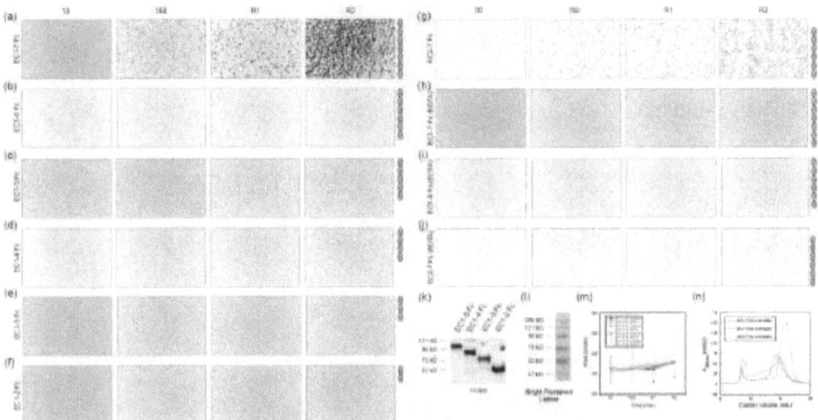

Figure 2.7 Bead aggregation assays for CDH17 fragments at various time points.
(a-j) Protein G beads coated with the full-length CDH17 EC1-7 Fc ectodomain (a,h), the C-terminal truncations EC1-6 Fc (b, i), EC1-5 Fc (c), EC1-4 Fc (d), EC1-3 Fc (e), EC1-2 Fc (f) and the N-terminal truncation EC2-7 Fc (g, j). Images show beads observed at the start of the experiment (T0), after 60 minutes (T60) followed by rocking for 1 min (R1) and 2 min (R2) in the presence of 2 mM $CaCl_2$ (a-g) or 2 mM EDTA (h-j). Bar – 500 μm. Panels shown for R2 timepoint in the presence of calcium and EDTA are also shown in Fig. 2.2 for EC1-7 Fc, EC1-6 Fc, and EC2-7 Fc. (k) Western blot shows expression and secretion of the protein fragments EC1-5 Fc, EC1-4 Fc, EC1-3 Fc and EC1-2 Fc. (l) Example of iBright Prestained ladder used for each western blot shown imaged under white light. (m) Aggregate size for the C-terminal truncation protein fragments EC1-5 Fc, EC1-4 Fc, EC1-3 Fc, and EC1-2 Fc at the start of the experiment (T0), after 60 min (T60) followed by rocking for 1 min (R1) and 2 min (R2). Error bars are standard error of the mean (n = 5 for all). (n) Elution profile for his-tagged *hs* CDH17 EC1-7, EC1-6 and EC2-7 on a GE Superose 6 10/300 column following native purification. Protein fragments are equivalent to the Fc-tagged fragments but were produced in Expi293F suspension cells. Aggregation peaks are small and all are comparable to the full-length EC1-7 His, suggesting that truncated proteins expressed from mammalian cells are properly folded and stable.

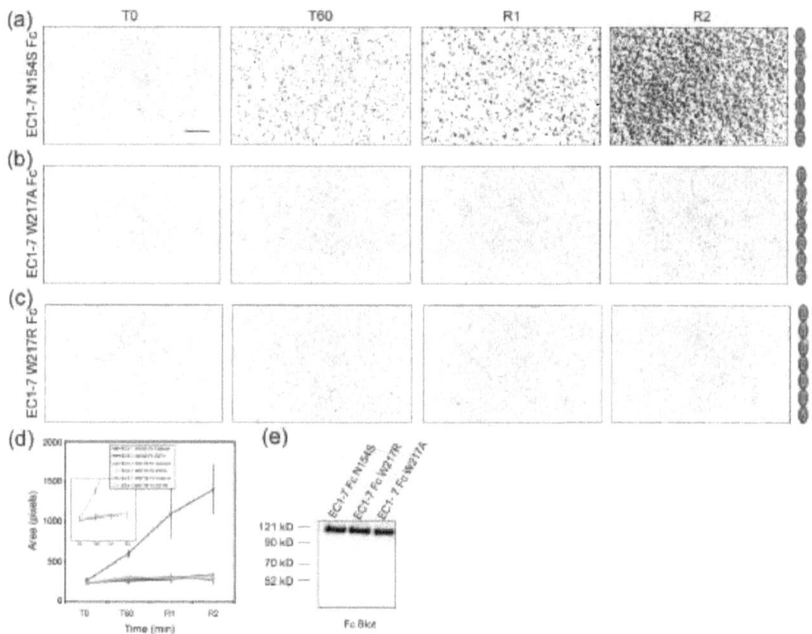

Figure 2.8 Bead aggregation assays of CDH17 mutants at various time points.
(a-c) Protein G beads coated with the full-length CDH17 EC1-7 Fc with mutations N154S (a), W217A (b), and W217R (c). Images show beads observed at the start of the experiment (T0), after 60 minutes (T60) followed by rocking for 1 min (R1) and 2 min (R2) in the presence of 2 mM $CaCl_2$. Bar – 500 μm. (d) Aggregate size for the full-length CDH17 EC1-7 Fc mutants at the start of the experiment (T0), after 60 min (T60) followed by rocking for 1 min (R1) and 2 min (R2). Inset shows zoom in of aggregate area up to 750 pixels. Error bars are standard error of the mean (n = 3 for all). (e) Western blot shows expression and secretion of the protein fragments *hs* CDH17 EC1-7 N154S Fc, EC1-7 W217A Fc, and EC1-7 W217R Fc.

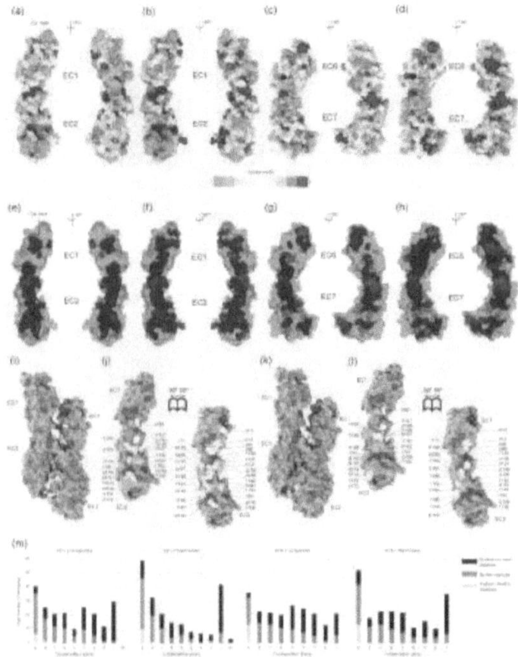

Figure 2.9 Sequence conservation in CDH17 EC1-2 and EC6-7 and residue surface exposure.
Sequence analyses of EC1-2 and EC6-7 were carried out using data from 234 species for CDH17 (a, c) and 113 mammalian species (b, d). (a) Surface representation of the *hs* CDH17 EC1-2 structure with residues colored according to conservation. Teal colors represent residues that are least conserved while dark magenta colors represent residues that are highly conserved. The suggested *cis* face is marked. (b) Surface representation of the *hs* CDH17 EC1-2 structure with residues colored according to conservation amongst mammalian species. Yellow surface represent residues for which Consurf could not calculate a conservation value. (c) Surface representation of *hs* CDH17 EC6-7 homology model with residues colored according to conservation as in (a). (d) Surface representation of *hs* CDH17 EC6-7 homology model with residues colored according to conservation as in (b). (e-f) Transparent surface representation of the *hs* CDH17 EC1-2 structure with most conserved residues amongst all species (e) and amongst mammals (f) shown in dark magenta surface. (g-h) Transparent surface representation of the *hs* CDH17 EC6-7 homology model with most conserved residues amongst all species (g) and amongst mammals (h) shown in dark magenta surface. (i-l) Surface representation of conservation of residues involved in the predicted *cis* interface from PISA for all species (i-j) and mammals (k-l). (j, l) Residues involved in the potential *cis* interface for all species (i) and mammals (k) are labeled for each chain. (m) Conservation score as a function of residue type for *hs* CDH17 EC1-2 and EC6-7 across all species or mammals. Each residue was classified as being surface exposed (> 20%), forming the hydrophobic core, or as involved in calcium binding and binned into their respective conservation score. A value of 9 represents a highly conserved residue, while a 1 represents residues that are not conserved. A value of N was given to residues for which Consurf was unable to determine conservation.

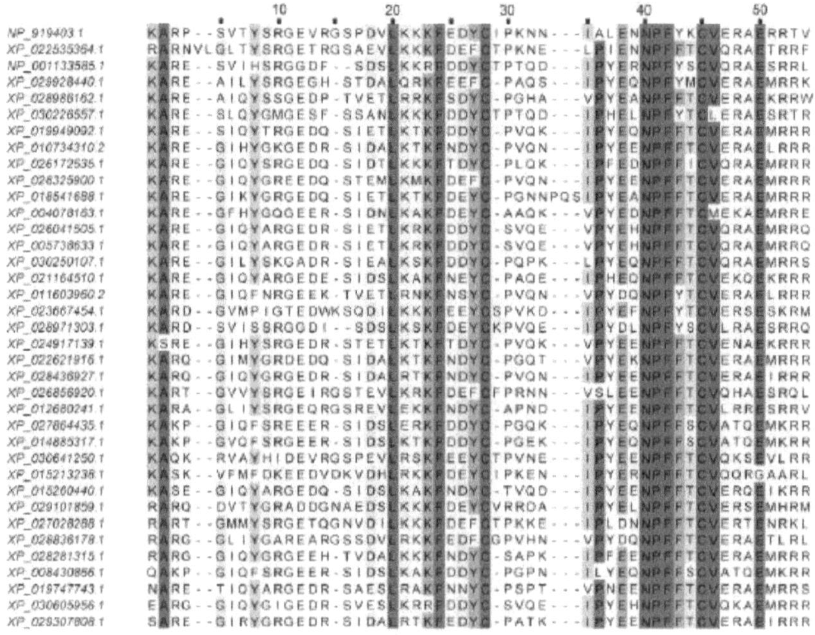

Figure 2.10 Sequence alignment of CDH17 EC2 insertion in fish species. Alignment is shown as in Fig. 2.4.

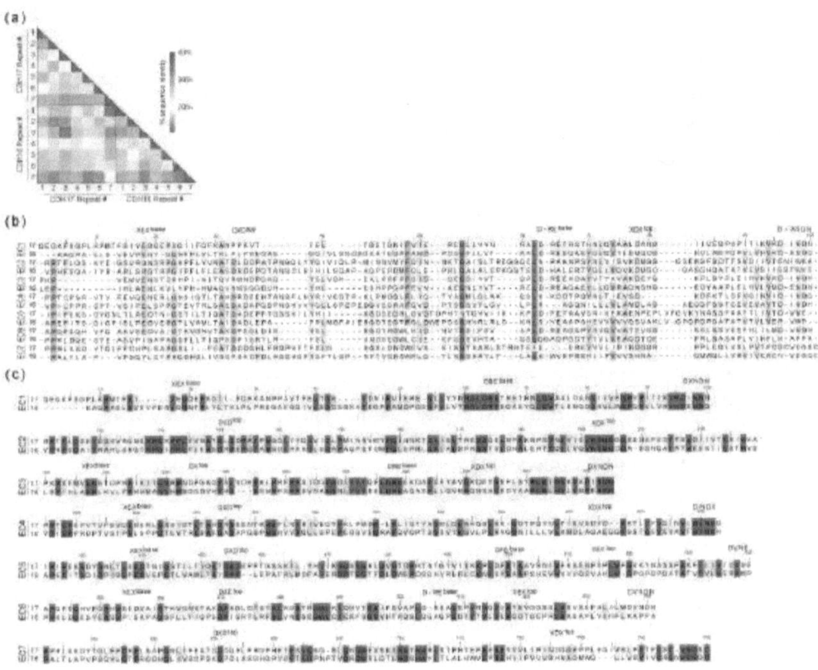

Figure 2.11 Sequence identity of CDH17 and CDH16.
(a) Heat map of percent identity between individual repeats of *hs* CDH17 and *hs* CDH16. (b) Sequence alignment of individual repeats of *hs* CDH17 and *hs* CDH16. Color denotes percent identity, with a threshold of 10%. Residue numbering corresponds to *hs* CDH17 EC1 and calcium-binding sites are denoted above the sequence. (c) Alignment of the EC repeats of *hs* CDH17 and *hs* CDH16. Color indicates percent identity, with darker blue indicating higher percent identity. Numbering corresponds to *hs* CDH17 and calcium-binding sites are denoted above the sequence.

Table 2.2 EC Alignment Sequences

Accession #	Species Name	Common Name
NP_004054.3	*Homo sapiens*	Human
XP_016815174.1	*Pan troglodytes*	Chimpanzee
NP_446429.1	*Rattus norvegicus*	Norway rat
NP_062727.1	*Mus musculus*	Mouse
XP_544179.3	*Canis lupus familiaris*	Dog
XP_011289700.1	*Felis catus*	Cat
XP_026920025.1	*Acinonyx jubatus*	Cheetah
NP_001092372.1	*Bos taurus*	Cow
XP_013852061.1	*Sus scrofa*	Pig
XP_022440252.1	*Delphinapterus leucas*	Beluga whale
NP_001186424.1	*Gallus gallus*	Chicken
XP_010403277.2	*Corvus cornix cornix*	Crow
XP_009273778.1	*Aptenodytes forsteri*	Emporer Penguin
XP_016848299.1	*Anolis carolinensis*	Green anole
XP_025055189.1	*Alligator sinensis*	Chinese alligator
NP_919403.1	*Danio rerio*	Zebrafish
XP_022535364.1	*Astyanax mexicanus*	Mexican tetra
NP_001133585.1	*Salmo salar*	Atlantic salmon
XP_007886135.1	*Callorhinchus milii*	Australian ghostshark
NP_001135580.1	*Xenopus tropicalis*	Tropical clawed frog

Table 2.3 Consurf Sequences

	Accession #	Species Name		Accession #	Species Name
1	NP_004054.3	*Homo sapiens*	118	XP_027748331.1	*Empidonax traillii*
2	NP_062727.1	*Mus musculus*	119	XP_026697595.1	*Athene cunicularia*
3	NP_446429.1	*Rattus norvegicus*	120	XP_025955628.1	*Dromaius novaehollandiae*
4	XP_016815174.1	*Pan troglodytes*	121	XP_023776536.1	*Cyanistes caeruleus*
5	NP_001092372.1	*Bos taurus*	122	XP_021398885.1	*Lonchura striata domestica*
6	XP_544179.3	*Canis lupus familiaris*	123	XP_021243790.1	*Numida meleagris*
7	XP_023504461.1	*Equus caballus*	124	XP_015711458.1	*Coturnix japonica*
8	XP_011289700.1	*Felis catus*	125	XP_015475257.1	*Parus major*
9	XP_003268363.1	*Nomascus leucogenys*	126	XP_014744163.1	*Sturnus vulgaris*
10	XP_002819325.1	*Pongo abelii*	127	XP_013797367.1	*Apteryx australis mantelli*
11	XP_017913717.1	*Capra hircus*	128	XP_013028003.1	*Anser cygnoides domesticus*
12	XP_005066250.1	*Mesocricetus auratus*	129	XP_011600627.1	*Aquila chrysaetos canadensis*
13	XP_004850163.1	*Heterocephalus glaber*	130	XP_010575449.1	*Haliaeetus leucocephalus*
14	XP_013852061.1	*Sus scrofa*	131	XP_010403277.2	*Corvus cornix cornix*
15	XP_004768625.1	*Mustela putorius furo*	132	XP_010200822.1	*Colius striatus*
16	XP_002759127.1	*Callithrix jacchus*	133	XP_010165060.1	*Antrostomus carolinensis*
17	XP_030718747.1	*Globicephala melas*	134	XP_010138961.1	*Buceros rhinoceros silvestris*
18	XP_029779387.1	*Suricata suricatta*	135	XP_010074240.1	*Pterocles gutturalis*
19	XP_029078795.1	*Monodon monoceros*	136	XP_010013536.1	*Nestor notabilis*
20	XP_028727471.1	*Peromyscus leucopus*	137	XP_010004805.1	*Chaetura pelagica*
21	XP_028625375.1	*Grammomys surdaster*	138	XP_009979398.1	*Tauraco erythrolophus*
22	XP_028389723.1	*Phyllostomus discolor*	139	XP_009944199.1	*Leptosomus discolor*
23	XP_027967117.1	*Eumetopias jubatus*	140	XP_009941846.1	*Opisthocomus hoazin*
24	XP_027802613.1	*Marmota flaviventris*	141	XP_009901267.1	*Picoides pubescens*
25	XP_027712793.1	*Vombatus ursinus*	142	XP_009890236.1	*Charadrius vociferus*
26	XP_027593127.1	*Pipra filicauda*	143	XP_009876351.1	*Apaloderma vittatum*
27	XP_027533018.1	*Neopelma chrysocephalum*	144	XP_009695763.1	*Cariama cristata*
28	XP_027514983.1	*Corapipo altera*	145	XP_009674315.1	*Struthio camelus australis*
29	XP_027462190.1	*Zalophus californianus*	146	XP_009636634.1	*Egretta garzetta*
30	XP_026987206.1	*Lagenorhynchus obliquidens*	147	XP_009467671.1	*Nipponia nippon*
31	XP_026351513.1	*Ursus arctos horribilis*	148	XP_009322727.1	*Pygoscelis adeliae*
32	XP_025870726.1	*Vulpes vulpes*	149	XP_009273778.1	*Aptenodytes forsteri*
33	XP_025778516.1	*Puma concolor*	150	XP_018767525.1	*Serinus canaria*
34	XP_025727274.1	*Callorhinus ursinus*	151	XP_008920371.2	*Manacus vitellinus*
35	XP_025289438.1	*Canis lupus dingo*	152	XP_008628963.1	*Corvus brachyrhynchos*
36	XP_025249753.1	*Theropithecus gelada*	153	XP_008500681.1	*Calypte anna*
37	XP_024621200.1	*Neophocaena asiaeorientalis asiaeorientalis*	154	XP_021136922.1	*Columba livia*
38	XP_024428081.1	*Desmodus rotundus*	155	XP_026653365.1	*Zonotrichia albicollis*

39	XP_023078634.1	*Piliocolobus tephrosceles*	156	XP_005445493.1	*Falco cherrug*
40	XP_022440252.1	*Delphinapterus leucas*	157	XP_005422979.1	*Geospiza fortis*
41	XP_021544134.1	*Neomonachus schauinslandi*	158	XP_012964449.2	*Anas platyrhynchos*
42	XP_021483221.1	*Meriones unguiculatus*	159	XP_003205227.1	*Meleagris gallopavo*
43	XP_021016727.1	*Mus caroli*	160	XP_030122283.1	*Taeniopygia guttata*
44	XP_020742321.1	*Odocoileus virginianus texanus*	161	XP_005151026.1	*Melopsittacus undulatus*
45	XP_019486198.1	*Hipposideros armiger*	162	XP_014820674.1	*Calidris pugnax*
46	XP_019383477.1	*Gavialis gangeticus*	163	XP_028592342.1	*Podarcis muralis*
47	XP_017702860.1	*Rhinopithecus bieti*	164	XP_026561790.1	*Pseudonaja textilis*
48	XP_017686075.1	*Lepidothrix coronata*	165	XP_026546647.1	*Notechis scutatus*
49	XP_017504401.1	*Manis javanica*	166	XP_024057359.2	*Terrapene carolina triunguis*
50	XP_017393595.1	*Cebus capucinus imitator*	167	XP_020640946.1	*Pogona vitticeps*
51	XP_016064589.1	*Miniopterus natalensis*	168	XP_019406068.1	*Crocodylus porosus*
52	>XP_015984995.1	*Rousettus aegyptiacus*	169	XP_015677347.1	*Protobothrops mucrosquamatus*
53	XP_015341220.1	*Marmota marmota marmota*	170	XP_015745132.1	*Python bivittatus*
54	XP_014697142.1	*Equus asinus*	171	XP_027674037.1	*Chelonia mydas*
55	XP_012628311.1	*Microcebus murinus*	172	XP_019351239.1	*Alligator mississippiensis*
56	XP_012499015.1	*Propithecus coquereli*	173	XP_006111073.1	*Pelodiscus sinensis*
57	XP_012294036.1	*Aotus nancymaae*	174	XP_025055189.1	*Alligator sinensis*
58	XP_011834032.1	*Mandrillus leucophaeus*	175	XP_023961073.1	*Chrysemys picta bellii*
59	XP_011792418.1	*Colobus angolensis palliatus*	176	XP_016848298.1	*Anolis carolinensis*
60	XP_011766526.1	*Macaca nemestrina*	177	NP_001135580.1	*Xenopus tropicalis*
61	XP_011382681.1	*Pteropus vampyrus*	178	NP_919403.1	*Danio rerio*
62	XP_010974309.1	*Camelus dromedarius*	179	XP_030250107.1	*Sparus aurata*
63	XP_010966693.1	*Camelus bactrianus*	180	NP_001133585.1	*Salmo salar*
64	XP_010831813.1	*Bison bison bison*	181	XP_007886135.1	*Callorhinchus milii*
65	XP_010617924.1	*Fukomys damarensis*	182	XP_022621916.1	*Seriola dumerili*
66	XP_010220026.1	*Tinamus guttatus*	183	XP_018541688.1	*Lates calcarifer*
67	XP_008846054.2	*Nannospalax galili*	184	XP_015805950.1	*Nothobranchius furzeri*
68	XP_008687366.1	*Ursus maritimus*	185	XP_030226557.1	*Gadus morhua*
69	XP_008590438.1	*Galeopterus variegatus*	186	XP_022535364.1	*Astyanax mexicanus*
70	XP_008536942.1	*Equus przewalskii*	187	XP_004078163.1	*Oryzias latipes*
71	XP_008146764.1	*Eptesicus fuscus*	188	XP_030641250.1	*Chanos chanos*
72	XP_008051725.1	*Carlito syrichta*	189	XP_030605956.1	*Archocentrus centrarchus*
73	XP_007999291.1	*Chlorocebus sabaeus*	190	XP_030014333.1	*Sphaeramia orbicularis*
74	XP_007537296.1	*Erinaceus europaeus*	191	XP_029958165.1	*Salarias fasciatus*
75	XP_007164981.1	*Balaenoptera acutorostrata scammoni*	192	XP_029928440.1	*Myripristis murdjan*
76	XP_023971681.1	*Physeter catodon*	193	XP_029360941.1	*Echeneis naucrates*
77	XP_007082836.1	*Panthera tigris altaica*	194	XP_029307808.1	*Cottoperca gobio*

78	XP_006975005.1	*Peromyscus maniculatus bairdii*	195	XP_028986162.1	*Betta splendens*
79	XP_006906876.1	*Pteropus alecto*	196	XP_028836178.1	*Denticeps clupeoides*
80	XP_006884689.1	*Elephantulus edwardii*	197	XP_028673585.1	*Erpetoichthys calabaricus*
81	XP_006859317.1	*Chrysochloris asiatica*	198	XP_028436927.1	*Perca flavescens*
82	XP_006729817.1	*Leptonychotes weddellii*	199	XP_028325900.1	*Gouania willdenowi*
83	XP_006209752.1	*Vicugna pacos*	200	XP_028281315.1	*Parambassis ranga*
84	XP_014420871.1	*Camelus ferus*	201	XP_027864435.1	*Xiphophorus couchianus*
85	XP_006161347.1	*Tupaia chinensis*	202	XP_027028288.1	*Tachysurus fulvidraco*
86	XP_006076349.1	*Bubalus bubalis*	203	XP_026856920.1	*Electrophorus electricus*
87	XP_005895592.1	*Bos mutus*	204	XP_026228820.1	*Anabas testudineus*
88	XP_005879292.1	*Myotis brandtii*	205	XP_026172535.1	*Mastacembelus armatus*
89	XP_005563762.1	*Macaca fascicularis*	206	XP_026041505.1	*Astatotilapia calliptera*
90	XP_005517458.1	*Pseudopodoces humilis*	207	XP_024124176.1	*Oryzias melastigma*
91	XP_005381861.1	*Chinchilla lanigera*	208	XP_023667454.1	*Paramormyrops kingsleyae*
92	XP_005362370.1	*Microtus ochrogaster*	209	XP_023256576.1	*Seriola lalandi dorsalis*
93	XP_005328711.2	*Ictidomys tridecemlineatus*	210	XP_020793171.1	*Boleophthalmus pectinirostris*
94	XP_005042355.1	*Ficedula albicollis*	211	XP_020462803.1	*Monopterus albus*
95	XP_004697772.1	*Echinops telfairi*	212	XP_019949092.1	*Paralichthys olivaceus*
96	XP_004679786.1	*Condylura cristata*	213	XP_019747743.1	*Hippocampus comes*
97	XP_004657087.1	*Jaculus jaculus*	214	XP_029101859.1	*Scleropages formosus*
98	XP_004597289.1	*Ochotona princeps*	215	XP_017280402.1	*Kryptolebias marmoratus*
99	XP_004474770.1	*Dasypus novemcinctus*	216	XP_015260440.1	*Cyprinodon variegatus*
100	XP_004413582.1	*Odobenus rosmarus divergens*	217	XP_014885317.1	*Poecilia latipinna*
101	XP_004275433.1	*Orcinus orca*	218	XP_014862605.1	*Poecilia mexicana*
102	XP_004047358.1	*Gorilla gorilla gorilla*	219	XP_013871075.1	*Austrofundulus limnaeus*
103	XP_010347701.1	*Saimiri boliviensis boliviensis*	220	XP_021164510.1	*Fundulus heteroclitus*
104	XP_021798192.1	*Papio anubis*	221	XP_012680241.1	*Clupea harengus*
105	XP_003821155.1	*Pan paniscus*	222	XP_028971303.1	*Esox lucius*
106	XP_003782523.1	*Otolemur garnettii*	223	XP_010778196.1	*Notothenia coriiceps*
107	XP_012397173.1	*Sarcophilus harrisii*	224	XP_010734310.2	*Larimichthys crocea*
108	XP_003507848.2	*Cricetulus griseus*	225	XP_008430866.1	*Poecilia reticulata*
109	XP_003479996.1	*Cavia porcellus*	226	XP_024917139.1	*Cynoglossus semilaevis*
110	XP_002710737.2	*Oryctolagus cuniculus*	227	XP_007556314.1	*Poecilia formosa*
111	XP_028917995.1	*Ornithorhynchus anatinus*	228	XP_015213238.1	*Lepisosteus oculatus*
112	XP_007488177.1	*Monodelphis domestica*	229	XP_014193984.1	*Haplochromis burtoni*
113	XP_004431226.2	*Ceratotherium simum simum*	230	XP_014328610.1	*Xiphophorus maculatus*
114	NP_001186424.1	*Gallus gallus*	231	XP_005738633.1	*Pundamilia nyererei*
115	XP_030799309.1	*Camarhynchus parvulus*	232	XP_024654805.1	*Maylandia zebra*
116	XP_030363898.1	*Strigops habroptila*	233	XP_011603960.2	*Takifugu rubripes*
117	XP_029867891.1	*Aquila chrysaetos chrysaetos*	234	XP_003450982.1	*Oreochromis niloticus*

Chapter 3. Species-dependent heterophilic and homophilic cadherin interactions in
intestinal intermicrovillar link1

Introduction

Enterocytes are specialized epithelial cells lining the luminal surface of the small intestine
and are fundamental players in nutrient absorption with a role in host defense [68,178–180].
Key to the aforementioned processes is the brush border, so named because the apical
surface of the enterocytes is coated by thousands of microvilli of similar length and size
organized in a hexagonal arrangement [181]. The organization and structure of the
microvilli is maintained by a network of cytoplasmic and transmembrane proteins
known as the intermicrovillar adhesion complex (IMAC), which includes proteins
with extracellular adhesive domains and cytoplasmic parts tethered to the actin
cytoskeleton that forms the interior of the microvilli [67,100–104]. Perturbations of
brush border function are often associated with disease [182–185] and, not surprisingly,
disruption of the IMAC components results in intestinal dysfunction [67,186,187].

The extracellular parts of the IMAC stem from a pair of cellular adhesion proteins that
belong to the cadherin superfamily, protocadherin-24 (PCDH24) and mucin-like
protocadherin (CDHR5) [56,67]. These two non-classical protocadherins form
intermicrovillar links essential for brush border morphogenesis and function
[67,100], stabilizing the microvillar hexagonal patterns as shown by freeze-etch electron
microscopy

and PCDH24 immuno-labeling combined with transmission electron microscopy [67]. Mutations that impair the PCDH24 and CDHR5 interaction, along with knockdowns of PCDH24 and CDHR5, cause a remarkable reduction in microvillar clustering *ex vivo* [67]. A PCDH24 knockout mouse model was found to be viable but body weight of mutant mice was lower than wild-type and there were defects in the packing of the microvilli in the brush border [103]. Despite the important physiological role played by PCDH24 and CDHR5, little is known about the molecular details of how these proteins interact to form the intermicrovillar links.

PCDH24 and CDHR5 are members of a large superfamily of proteins with extracellular domains that often mediate adhesion and that have contiguous and similar, but not identical, extracellular cadherin (EC) repeats. The linker regions between the EC repeats typically bind three calcium ions essential for adhesive function [1,188–190]. PCDH24 belongs to the Cr-2 subfamily and has 9 EC repeats, a membrane adjacent domain (MAD10), a transmembrane domain, and a C-terminal cytoplasmic domain [56,105–108,191]. CDHR5 belongs to the Cr-3 subfamily and has 4 EC repeats, a mucin-like domain (MLD), a transmembrane domain, and a C-terminal cytoplasmic domain. Some human CDHR5 isoforms lack the MLD, which is not essential for the interaction with human PCDH24 [67]. Sequence alignments of the N-terminal repeats (EC1-3) suggest that PCDH24 and CDHR5 are similar to another pair of heterophilic interacting cadherins: cadherin 23 (CDH23) and protocadherin 15 (PCDH15), the cadherins responsible for the formation of inner-ear tip links [7,33,65,109,110]. PCDH24 is most similar to CDH23, having the residues and elongated

N-terminal β-strand that are predicted to favor the formation of an atypical calcium-binding site at its tip [109,110]. CDHR5 is most similar to PCDH15, and features cysteine residues predicted to form a disulfide bond at its tip [7,33]. In addition, a mutation in CDHR5, R84G (residue numbering throughout the text corresponds to processed proteins, see Materials and Methods), which mimics a deafness-related mutation in PCDH15, interferes with the intermicrovillar links formed by PCDH24 and CDHR5 [67], suggesting that these proteins interact in a similar fashion to the heterophilic tip-link "handshake" used by CDH23 and PCDH15 [7,33,192].

To better understand how PCDH24 and CDHR5 interact to form intermicrovillar links, we used a combination of structural and bead aggregation assays to characterize the adhesive properties and mechanism of these cadherin family members. We present the X-ray crystallographic structures of *Homo sapiens* (*hs*) PCDH24 EC1-2, *Mus musculus* (*mm*) PCDH24 EC1-3, and *hs* CDHR5 EC1-2 refined at 2.3 Å, 2.1 Å, and 1.9 Å resolution, respectively. The structures give insight into possible binding mechanisms among species, which are further probed by bead aggregation assays revealing that the human and mouse intermicrovillar cadherins do not engage in the same homophilic and heterophilic interactions. Based on these results, we suggest models for the PCDH24/CDHR5 heterophilic interaction utilized to form intermicrovillar links.

Results

PCDH24 and CDHR5 tip sequences are poorly conserved across species

Intermicrovillar links formed by PCDH24 and CDHR5 in the enterocyte brush border are similar to inner-ear tip links formed by CDH23 and PCDH15. Both types of links are essential for the development, assembly, and function of actin-based structures (microvilli in the gut and stereocilia in the inner ear), and both are made of long cadherin proteins that have similar cytoplasmic partners involved in interactions with the cytoskeleton and in signaling [100,114]. Sequence analyses have revealed similarities between the tips of PCDH24 and CDH23, and between the tips of CDHR5 and PCDH15 [7]. Given that CDH23 and PCDH15 engage in a heterophilic "handshake" complex involving their EC1-2 tips, it has been proposed that PCDH24 and CDHR5 might use a similar binding mechanism [7,67]. To further explore this hypothesis, we performed sequence analyses comparing the tips of these cadherins to each other, to classical cadherins, and across species.

Interestingly, alignments of EC repeat sequences across up to 20 species for intermicrovillar and tip-link cadherins reveal poor conservation for PCDH24 and CDHR5 when compared to CDH23 and PCDH15. Average percent identity is 51.6% for CDH23 EC repeats and 45.5% for PCDH15 repeats, with the N- and C-terminal ends being more conserved than the middle EC repeats [24,36]. In contrast, average percent identity is 16.4% for PCDH24 and 11% for CDHR5, considerably lower when compared to values for CDH23 and PCDH15. Moreover, the N- and C-terminal ends of PCDH24 and CDHR5 tend to be less conserved than the middle region of these proteins (Figure 3.1A, 3.8, and

3.9; Tables 3.2, 3.3, 3.4, 3.5, and 3.6). Nevertheless, multiple sequence alignments of PCDH24 and CDH23 EC1 repeats confirm that PCDH24 has an elongated N-terminus that should facilitate the formation of calcium-binding site 0 as observed in structures of CDH23 (Figure 3.10A). Binding of calcium at this site in CDH23 is mediated by several acidic residues, which are also present and conserved in PCDH24. Similarly, sequence alignments of CDHR5 and PCDH15 EC1 repeats confirm that CDHR5 has the conserved cysteine residues that form a stabilizing disulfide bond at the tip of PCDH15 EC1 (Figure 3.10B). None of these proteins feature the tryptophan residues that mediate homophilic binding in classical cadherins [8,38,146], supporting the hypothesis that PCDH24 and CDHR5 might form a tip-link-like handshake complex mediating adhesion.

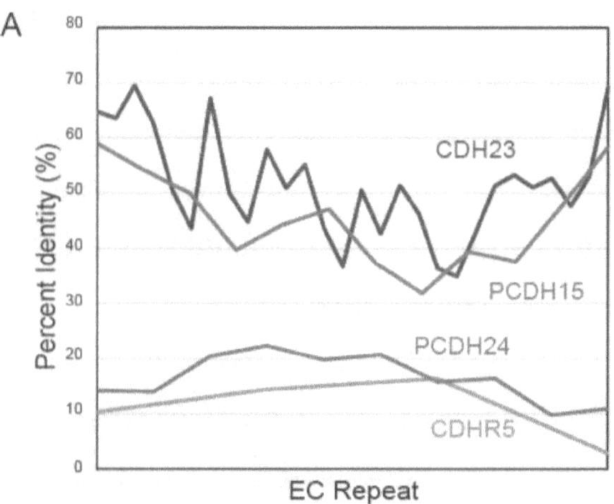

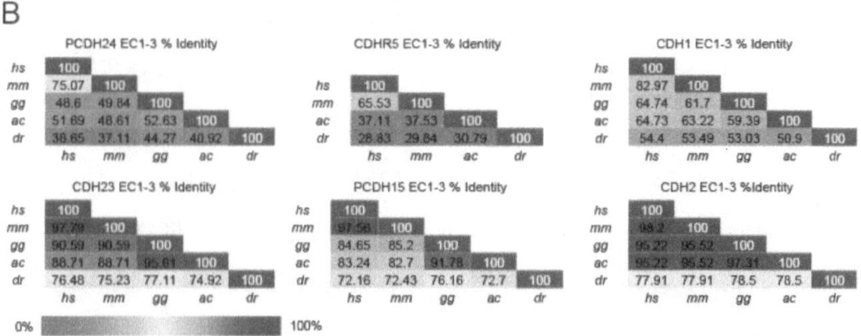

Figure 3.1 Comparison of sequence conservation across species.
(A) Percent identity across the extracellular domains of PCDH24 (EC1-MAD10), CDH23 (EC1-MAD28), CDHR5 (EC1-4), and PCDH15 (EC1-MAD12) plotted against EC repeat numbers. Overall the cadherins of the inner-ear tip link, CDH23 and PCDH15, have higher identity across a variety of species while the intermicrovillar-link cadherins, PCDH24 and CDHR5, have poor sequence conservation across species (Figure 3.8 and 3.9; Tables 3.3, 3.4, 3.5 and 3.6). N- and C-termini are more conserved in CDH23 and PCDH15 than the middle region of these proteins. An opposite trend is observed for PCDH24 and CDHR5. (B) Percent identity of the first three EC repeats of PCDH24, CDHR5, CDH23, PCDH15, CDH1 and CDH2 demonstrate that PCDH24 and CDHR5 sequences are not highly conserved across several species (*Homo sapiens, hs*; *Mus musculus, mm*; *Gallus gallus, gg*; *Anolis carolinesis, ac*; *Danio rerio, dr*). However, the N-terminal repeats of CDH23, PCDH15, CDH1, and CDH2 are highly conserved. Sequences were obtained from NCBI (Table 3.7 and Materials and Methods), except for *gg* CDHR5, which is unavailable.

Pairwise sequence identity computed across up to five species for the N-terminal repeats

(EC1-3) of PCDH24, CDHR5, CDH23, PCDH15, and the classical cadherins E-cadherin

(CDH1) and N-cadherin (CDH2), also show different trends of sequence conservation

(Figure 3.1B and Table 3.7). Average percent identity in pairwise comparisons is high

for CDH2 (89%), CDH23 (86%), and PCDH15 (82%), moderate for CDH1 (61%) and

lowest for PCDH24 (49%) and CDHR5 (38%). This is evident when comparing

sequences from the most evolutionary distant species (human and fish). For instance, *hs*

and *Danio rerio (dr)* PCDH24 EC1-3 sequences are 39% identical, while *hs* and *dr*

CDH1 EC1-3 sequences are 54% identical. Sequence differences are still large for the

human and mouse PCDH24 and CDHR5 protein tips, with 75% identity for the *hs* an *mm*

PCDH24 EC1-3 sequences and 66% identity for the *hs* and *mm* CDHR5 EC1-3

sequences (compared to CDH1, CDH2, CDH23, and PCDH15 EC1-3 sequences with *hs*

an *mm* pairwise identities between 83% and 98%) (Figure 3.1B). Proteins with sequence

identity as low as 30% can share similar folds [193], yet the low PCDH24 and CDHR5

sequence conservation in multiple sequence alignments and pair-wise comparisons

suggests that details of their structures and function differ across species.

**Structures of human and mouse PCDH24 tips reveal N-terminal conformational
variability**

Previous bead aggregation assays using the *hs* PCDH24 protein revealed that its full-length

extracellular domain can mediate adhesion through homophilic interactions with itself and

through heterophilic interactions with *hs* CDHR5 [67]. Members of the cadherin superfamily

of proteins often engage in homophilic and heterophilic interactions using their N-terminal tips (EC1 for classical cadherins, EC1-2 for tip-link cadherins, and EC1-4 for clustered, δ1, and δ2 protocadherins) [18,22,29,31–33,38,51,194,195]. Therefore, to gain insights into the molecular basis of PCDH24-mediated adhesion and to explore the consequences of poor sequence conservation across species, we solved the crystal structures of *hs* PCDH24 EC1-2 and *mm* PCDH24 EC1-3 refined at 2.3 Å and 2.1 Å resolution, respectively (Table 3.1). These structures, obtained from crystals of proteins expressed in *E. coli* and refolded from inclusion bodies (see Materials and Methods), revealed both expected and unexpected architectural features.

Table 3.1 Statistics for PCDH24 and CDHR5 structures

Data Collection and Refinement	*Hs* PCDH24 EC1-2	*Mm* PCDH24 EC1-3	*Hs* CDHR5 EC1-2
Space Group	$P\,2_1$	$P\,3\,2\,1$	$I\,2\,2\,2$
Unit cell parameters			
a, b, c (Å)	270.9, 119.8, 74.3	74.8, 74.8, 140.0	50.4, 78.4, 121.4
α, β, γ (°)	90, 104.12, 90	90, 90, 120	90, 90, 90
Molecules per asymmetric unit	4	1	1
Beam Source	APS 24-ID-C	APS 24-ID-C	APS 24-ID-E
Wavelength (Å)	0.97910	0.97910	0.97918
Resolution limit (Å)	2.3	2.1	1.9
Unique Reflections	50945 (2128)	27040 (1237)	19398 (943)
Redundancy	3.1 (2.6)	8.60 (5.20)	24.10 (21.80)
Completeness (%)	95.5 (80.3)	99.2 (90.6)	99.7 (99.7)
Average I/σ (I)	10.9 (3)	27 (4.4)	46.14 (14.50)
R_{merge}	0.105 (0.255)	0.078 (0.337)	0.074 (0.263)
Refinement			
Resolution range (Å)	50 – 2.3 (2.34 – 2.3)	50 - 2.1 (2.14 – 2.1)	50-1.9 (1.93-1.9)
Residues (atoms)	861 (6610)	318 (2447)	204 (1656)
Water Molecules	238	98	128
R_{work} (%)	20.5 (27.8)	21.0 (34.3)	19.7 (20.9)
R_{free} (%)	23.5 (33.0)	24.3 (39.6)	23.4 (25.9)
Rms deviations			
Bond lengths (Å)	0.013	0.014	0.013
Bond angles (°)	1.460	1.632	1.652
B factor average			
Protein	41.00	64.31	30.53
Ligand/ion	36.90	51.49	39.42
Water	31.65	50.99	33.6
Ramachandran plot regions			
Most favored (%)	91.5	90.7	91.2
Additionally allowed (%)	8.3	8.9	8.8
Generously allowed (%)	0.3	0.4	0.0
Disallowed (%)	0.0	0.0	0.0
PDB ID code	5CZR	5CYX	6OAE

The *hs* PCDH24 EC1-2 structure has four molecules in the asymmetric unit including residues N1 to P215 for each protomer, while the *mm* PCDH24 EC1-3 structure has one molecule in the asymmetric unit including residues N1 to T324. Both structures had good-quality electron density maps that allowed the unambiguous positioning of side chains for most of the residues (Figure 3.11A-B). As expected, all EC repeats in the PCDH24 structures adopt the typical seven β-strand Greek-key fold seen in extracellular domains of cadherin proteins, with β-strands labeled A to G for each repeat (Figure 3.2A-D). Both *hs* and *mm* PCDH24 EC2 repeats have extended F and G strands, which are linked by a disulfide bond (S187:S201) that stabilizes an unusually long FG β-hairpin near the EC1-2 linker (Figure 3.2A-D and 3.11A).

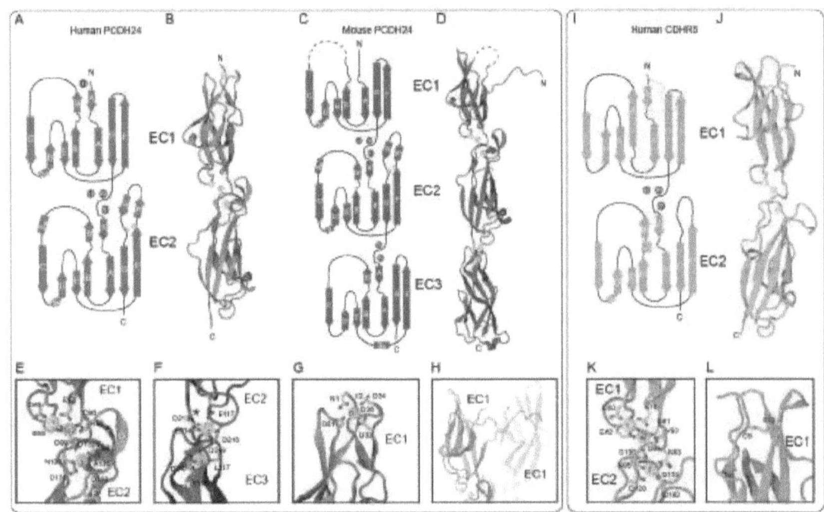

Figure 3.2 Structures of human (*hs*) PCDH24 EC1-2, mouse (*mm*) PCDH24 EC1-3, and human (*hs*) CDHR5 EC1-2.

(A) Topology of *hs* PCDH24 EC1-2. The EC1 and EC2 repeats have a typical cadherin fold with 7 β-strands. A disulfide bond in EC2 is highlighted in orange. Calcium ions are shown as green circles. (B) A ribbon representation of *hs* PCDH24 EC1-2 with calcium ions in green. (C) Topology of the *mm* PCDH24 EC1-3 shown as in (A). The dashed line indicates a loop not resolved. (D) A ribbon representation of *mm* PCDH24 EC1-3 as in (B). The structure is similar to *hs* PCDH24 EC1-2, but the N-terminus projects away from EC1. A N-terminal calcium ion was not present. (E) Detail of the linker between EC1 and EC2 of *hs* PCDH24 showing canonical calcium-binding sites. (F) Detail of the *mm* PCDH24 EC2-3 linker, which lacks several canonical calcium-binding residues and thus only has two calcium ions bound. Shown as in (C). Some sidechain and backbone atoms are not shown for clarity. An asterisk (*) indicates the position where a calcium residue would be found in the canonical linker. (G) Detail of the N-terminal calcium bound at the tip of EC1 in the *hs* PCDH24 EC1-2 structure. A similar calcium-binding motif is seen at the tip of CDH23. Residues coordinating calcium are shown in stick with backbone atoms omitted. (H) Detail of the N-terminus of *mm* PCDH24 EC1-3 (blue) showing its interaction with another protomer (cyan) in the asymmetric unit. (I) Topology of *hs* CDHR5 EC1-2 shown as in (A). (J) A ribbon representation of *hs* CDHR5 EC1-2 with calcium ions in green. (K) Detail of the linker between EC1 and EC2 of *hs* CDHR5 showing canonical calcium-binding sites. Shown as in (I-J). (L) Detail of the disulfide bond in EC1 in the *hs* CDHR5 EC1-2 structure. A similar disulfide bond is seen in PCDH15 EC1.

The PCDH24 EC1-2 linker region is canonical in both human and mouse structures and features calcium ions bound to sites 1, 2, and 3, which are coordinated by acidic residues from the cadherin motif NTerm-XEX-DXD-D(R/Y)(D/E)-XDX-DXNDN-CTerm (Figure 3.2A-E and 3.12). Interestingly, the structure of mouse PCDH24 has a non-canonical calcium-binding linker between repeats EC2 and EC3, with only two bound calcium ions and a straight conformation (Figure 3.2C-D, F and 3.12). This linker lacks necessary conserved residues to bind the calcium ion that would occupy site 1 as the canonical DXE sequence is replaced with residues 172SYN174. Although the canonical DXNDN sequence motif is 213DXPDL217 at the EC2-3 linker, typical coordination of calcium at site 3 by the asparagine carboxamide oxygen is replaced by coordination through the backbone carbonyl group of a glutamate residue in our structure (E214 in mouse, Q214 in human). The last residue of the motif (typically asparagine, replaced here by L217) coordinates calcium at site 3 through its backbone carbonyl group, an interaction that is unchanged by residue variability at this position. Sequence conservation across species for the DXE/SYN and DXNDN/DXPDL motifs suggest that most species have a non-canonical linker between repeats EC2 and EC3 with two bound calcium ions as observed in our *mm* PCDH24 EC1-3 structure. Equilibrium and steered molecular dynamics (SMD) simulations of *mm* PCDH24 EC1-3 show that rigidity and mechanical strength are not compromised in this non-canonical linker (Figure 3.13).

As expected for members of the Cr-2 family of non-clustered protocadherins, the *hs* PCDH24 EC1-2 structure shows an occupied site 0 calcium-binding site at the tip of EC1, with calcium-coordinating residues similar to those observed in structures of CDH23 (1N, 32DXDXD36, 80XDX^{TOP}; Figure 3.2A-B, G, and 3.14A). Unexpectedly, the *mm* PCDH24 EC1-3 structure shows an unoccupied site 0, with an extended N-terminus protruding away from EC1 as observed for classical cadherins that exchange their N-terminal strands to form adhesive bonds (Figure 3.2C-D, H, and 3.14B). The sequence motifs involved in calcium coordination at site 0 are conserved, suggesting that crystallization conditions, including 3M LiCl, facilitated the opening of this site. Alternatively, other differences between mouse and human sequences might be responsible for the structural divergence. Most notably, residue R86 near the FG loop and the XDX^{TOP} motif of *mm* PCDH24 EC1 seems to be incompatible with a closed N-terminal conformation and an occupied calcium-binding site 0. The open and closed conformations observed for PCDH24 EC1 will likely determine the type of homophilic or heterophilic interactions that mediate adhesion.

Crystal contacts in human and mouse PCDH24 structures suggest distinct adhesive interfaces

Crystallographic contacts observed in cadherin structures have revealed multiple physiologically relevant interfaces [22,29,31,32,150]. The crystallographic packing observed for the *hs* and *mm* PCDH24 structures shows various interfaces that further highlight differences across species. The asymmetric unit of the *hs* PCDH24 EC1-2 structure shows 4 molecules with two similar anti-parallel *trans* dimers (Figure 3.15A) arranged perpendicular to one another to form a dimer of dimers (Figure 3.15B). Each antiparallel dimer positions EC1 from one protomer in front of EC2 from the next protomer, slightly shifted with respect to each other and with the extended FG β-hairpins near the EC1-2 linker regions mediating the interaction head to head. The antiparallel dimers have interface areas of 979.2 $Å^2$ and 938.6 $Å^2$ for protomers A:D and B:C, respectively. These values are both larger than an empirical threshold (856 $Å^2$) that distinguishes biologically relevant interactions from crystallographic packing artifacts [148]. Two other interfaces are small and unlikely to be biologically relevant (Figure 3.15C-D). Previous data has shown that *hs* PCDH24 mediates homophilic *trans* adhesion when *hs* CDHR5 is not present [67]. It is possible that the antiparallel EC1-2 interface facilitates homophilic adhesion mediated by *hs* PCDH24, yet our SMD simulations suggest that this interface is weak (Figure 3.16).

In contrast to the crystal contacts observed in the *hs* PCDH24 EC1-2 structure, the crystal packing of *mm* PCDH24 EC1-3 (space group *P*321) results in multiple large interfaces generated from rotations of a single molecule in the asymmetric unit. A parallel trimer is present around the axis of symmetry, with the EC1 N-termini from each protomer protruding away and inserting into pockets of adjacent EC1 repeats in a clockwise fashion, thereby interlocking the EC1 domains (Figure 3.17A-B). This is possible because of the open conformation observed for the mouse N-terminus of EC1 without a bound calcium ion at site 0, and it is reminiscent of how the EC1 N-terminus of classical cadherins is exchanged resulting in the insertion of W2 into an hydrophobic binding pocket of the partner molecule (Figure 3.14C-D) [38,146]. Interestingly, a glycosylation site is predicted at N9, which might interfere with or regulate the formation of interlocking EC1 repeats, as has been observed for other cadherins [150].

The interlocking of mouse EC1 repeats in the trimer facilitates the formation of two parallel *cis* interfaces between EC1-3 protomers involving the FG β-hairpin interacting with the neighboring EC2-3 linker. These *cis* interfaces are large with an interface area of 1352.6 Å² between two monomers (Figure 3.17A). In addition, a fully overlapped antiparallel *trans* interface is observed between EC1-3 protomers with an area of 1221.2 Å² (Figure 3.17C). Sugars at a second predicted glycosylation site located in the center of the *cis* trimer (p.N161), and at a third predicted glycosylation site near the *trans* interface between EC1 and EC3 (p.N280), may interfere with or regulate the formation of these interfaces. Together, the *cis* trimer and *trans* dimer could form a glycosylation-modulated interlocking arrangement of PCDH24 molecules that is different from the *trans* interaction in the *hs* PCDH24 EC1-2 structure. Sequence conservation mapped to the *mm* EC1-3 structure reveals that only core residues are highly conserved, highlighting poor conservation for surface exposed residues potentially involved in adhesive interactions (Figure 3.18). Motivated by the sequence and structural differences observed for the human and mouse PCDH24 tips, we used bead aggregation assays to determine the minimum adhesive unit of PCDH24 homophilic binding in both species.

Human, but not mouse PCDH24 mediates homophilic adhesion

To determine which EC repeats are essential for *hs* and *mm* PCDH24 homophilic adhesion we created a library of constructs encoding for the full-length extracellular domains (EC1-MAD10) and various truncations (EC1-7, EC1-6, EC1-5, EC1-4, EC1-3, EC1-2, and EC1), all fused to a C-terminal Fc tag. These protein fragments were produced in HEK293T cells and used for aggregation assays with Protein G magnetic beads (see Materials and Methods).

As expected, bead aggregation experiments using *hs* PCDH24 EC1-MAD10Fc showed large calcium-dependent aggregates (Figure 3.3A, L-M, and 3.22A, I). Similar large calcium-dependent aggregates were seen for *hs* PCDH24 EC1-7Fc through EC1-3Fc (Figure 3.3B-F, L-M, and 3.22B-F). Further truncations indicate that smaller *hs* PCDH24 fragments, including just EC1, are capable of facilitating the formation of small bead aggregates that increase in size with rocking but never reach the larger sizes observed when using fragments with 3 or more EC repeats (Figure 3.3G-H, K-M, and 3.22G-H). Controls performed using EDTA showed that constructs are unable to aggregate when calcium is depleted, except when using EC1Fc, possibly stabilized by the adjacent Fc-tag (Figure 3.3I-M, and 3.22I). These results suggest that *hs* PCDH24 mediates calcium-dependent homophilic adhesion, with at least three distinct modes of adhesion, two calcium-dependent mediated by *hs* PCDH24 EC1-3 and PCDH24 EC1-2, and another one calcium-independent mediated by *hs* PCDH24 EC1 alone. It is possible that each of these modes uses a different interface, and that the antiparallel interface observed in our *hs* PCDH24 EC1-2 structure is a low-affinity intermediate state that facilitates further interdigitation for the adhesive mode that drives the formation of larger bead aggregates.

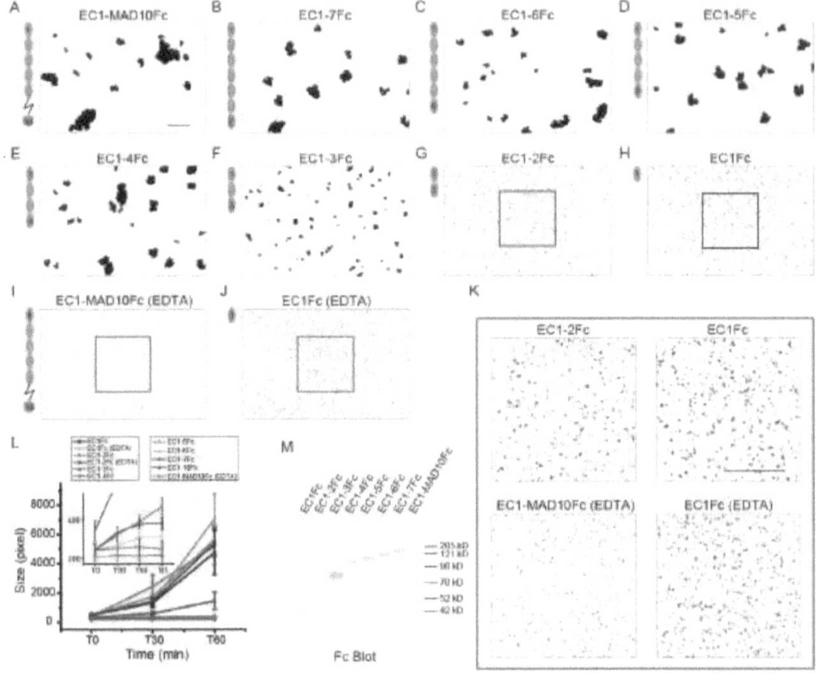

Figure 3.3 Homophilic binding assays of hs PCDH24.
(A-H) Protein G beads coated with the full-length hs PCDH24 extracellular domain (A) and its C-terminal truncation versions (B-H). Each image shows bead aggregation observed after 60 min in the presence of 2 mM CaCl$_2$. (I-J) Protein G beads coated with the full-length hs PCDH24 extracellular domain (I) and EC1 (J) in the presence of 2 mM EDTA, shown as in (A-H). (K) Detail of binding assay results for EC1-2Fc, EC1Fc, EC1-MAD10Fc with EDTA, and EC1Fc with EDTA. Aggregation is still present for the shortest constructs. Bars – 500 µm. (L) Aggregate size for full-length and truncated versions of hs PCDH24 extracellular domains at the start of the experiment (T0), and after 30 and 60 min (T30, T60). Inset shows aggregate size of the shortest constructs compared to the EDTA control of EC1-MAD10Fc. An additional one-minute rocking step is denoted by R1. Error bars are standard error of the mean (n indicated in Table 3.8). (M) Western blot shows expression and secretion of full-length and truncated versions of hs PCDH24.

Given the low sequence identity between the human and mouse N-terminal EC repeats of PCDH24 when compared to CDH23, and the structural differences observed between *hs* PCDH24 EC1-2 and *mm* PCDH24 EC1-3, bead aggregation assays were also performed with the full-length extracellular domain of *mm* PCDH24. Bead aggregation was not observed for *mm* PCDH24 EC1-MAD10Fc (Figure 3.4A-B, E, G). These results suggest that the large crystallographic antiparallel *trans* interface formed by PCDH24 EC1-3 (Figure 3.17C) does not mediate bead aggregation under the conditions tested and reveal that PCDH24 mediated homophilic adhesion is species-dependent.

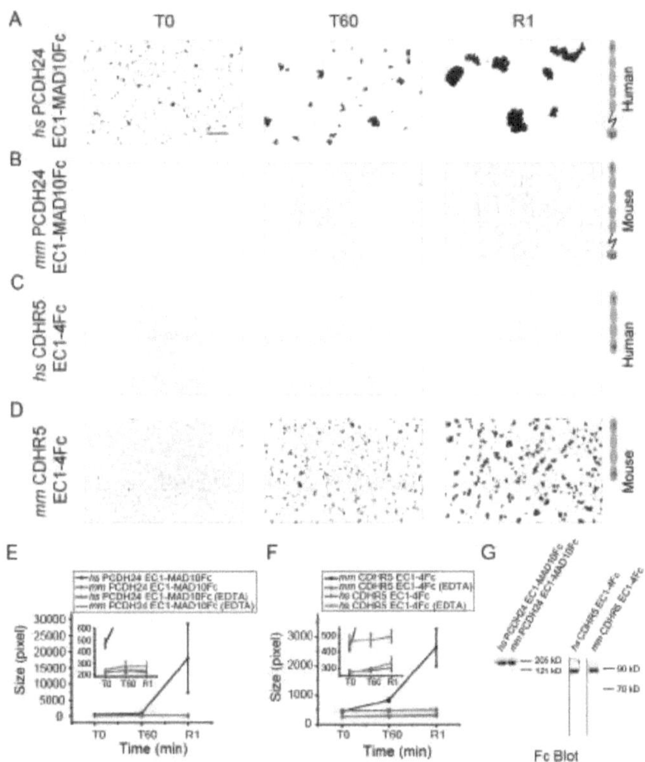

Figure 3.4 Comparison of homophilic binding assays of hs and mm PCDH24 and CDHR5.
(A-D) Protein G beads coated with full-length hs PCDH24 (A), full-length mm PCDH24 (B), full-length hs CDHR5 (C), and full-length mm CDHR5 (D) cadherin extracellular domains, including MAD10 for PCDH24. None of the CDHR5 fragments included the mucin-like domain. Images show the aggregation observed at the start of the experiment (T0), after 60 min (T60) followed by rocking for 1 min (R1) in the presence of 2 mM CaCl₂. Bar - 500 μm. (E) Aggregate size for full-length hs and mm PCDH24 extracellular domains in the presence of 2 mM CaCl₂ and 2 mM EDTA at the start of the experiment (T0), after 60 min (T60) followed by rocking for 1 minute (R1). Inset shows aggregate size of full-length mm PCDH24 extracellular domain compared to the EDTA control of full-length hs and mm PCDH24 extracellular domains. (F) Aggregate size for full-length hs and mm CDHR5 cadherin extracellular domains in the presence of 2 mM CaCl₂ and 2 mM EDTA at the start of the experiment (T0), after 60 min (T60) followed by rocking for 1 min (R1). Inset shows aggregate size of the full-length hs CDHR5 cadherin extracellular domain compared to the EDTA control of full-length hs and mm CDHR5 cadherin extracellular domain. Error bars in (E) and (F) are standard error of the mean (n indicated in Table 3.8). (G) Western blot shows expression and secretion of full-length hs and mm PCDH24 and CDHR5 cadherin extracellular domains.

Mouse, but not human CDHR5 mediates homophilic adhesion

The full-length extracellular domain of *hs* CDHR5 was previously shown to not mediate homophilic adhesion [67]. The low sequence identity between the human and the mouse CDHR5, especially at their N-termini, and the species-dependent homophilic adhesive behavior of PCDH24, prompted us to test if *mm* CDHR5 could mediate adhesion. Full-length cadherin extracellular domains (without mucin-like domains) of both human and mouse CDHR5 were used for bead aggregation assays. As expected, *hs* CDHR5 EC1-4Fc did not mediate bead aggregation in the presence and the absence of calcium. However, *mm* CDHR5 EC1-4Fc did mediate bead aggregation in the presence of calcium (Figure 3.4C-D, F-G). These results suggest that, like PCDH24, CDHR5 homophilic adhesiveness is species-dependent.

To determine the minimal adhesive unit of *mm* CDHR5, a library of truncated cadherin fragments was generated and used for bead aggregation assays. Calcium-dependent bead aggregation was observed when using beads coated with *mm* CDHR5 EC1-4Fc and EC1-3Fc, but not when using EC1-2Fc and EC1Fc (Figure 3.5A-D, G-I, and 3.23A-E). Hence, *mm* CDHR5 EC1-3 is sufficient to mediate homophilic adhesion.

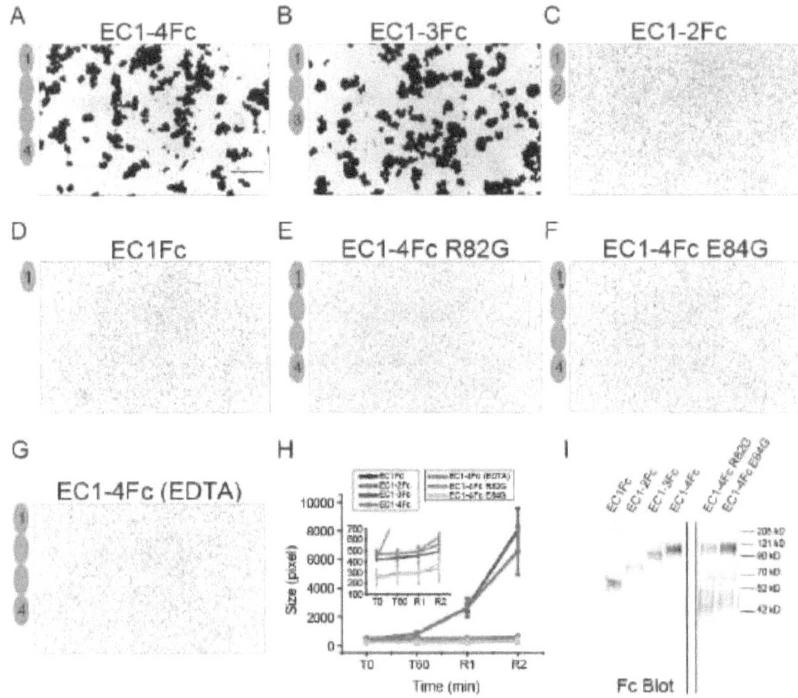

Figure 3.5 Homophilic binding assays of mm CDHR5.
(A-F) Protein G beads coated with the full-length mm CDHR5 cadherin extracellular domain (A) and its C-terminal truncation versions (B-D). Each image shows bead aggregation observed after 60 min followed by rocking for 2 min in the presence of 2 mM CaCl$_2$. (E-F) Protein G beads coated with mutants mm CDHR5 EC1-4Fc R82G (E) and EC1-4Fc E84G (F) observed after 60 min followed by rocking for 2 min in the presence of 2 mM CaCl$_2$. (G) Protein G beads coated with the full-length mm CDHR5 cadherin extracellular domain in the presence of 2 mM EDTA shown as in (A). Bar - 500 µm. (H) Aggregate size for full-length, truncated versions and mutants of mm CDHR5 cadherin extracellular domains at the start of the experiment (T0), after 60 min (T60) followed by rocking for 1 min (R1) and 2 min (R2). Inset shows aggregate size of EC1Fc, EC1-2Fc, the EDTA control of EC1-4Fc and EC1-4Fc mutants R82G and E84G. Error bars are standard error of the mean (n indicated in Table 3.8). (I) Western blot shows expression and secretion of full-length, truncated, and mutant versions of mm CDHR5 cadherin extracellular domains.

To provide insight into the interface used by *mm* CDHR5 to mediate homophilic adhesion, we tested mutations that alter a site similar to that mutated in *hs* CDHR5 (R84G) preventing heterophilic binding to *hs* PCDH24. Bead aggregation assays were performed with both *mm* CDHR5 EC1-4Fc R82G and E84G, and both mutants abolished bead aggregation of *mm* CDHR5 EC1-4Fc in the presence of calcium (Figure 3.5E-F, H-I). These results suggest that the interface used for mediating heterophilic adhesion between PCDH24 and CDHR5 and the interface used for mediating homophilic adhesion by *mm* CDHR5 might be similar.

Structure of human CDHR5 EC1-2 reveals a PCDH15-like fold and suggests adhesive interfaces

To gain insight into CDHR5's adhesive function we solved the crystal structure of *hs* CDHR5 EC1-2 refined at 1.9 Å resolution (Table 3.1). This structure, obtained using protein expressed in *E. coli* and refolded from inclusion bodies (see Materials and Methods), had one molecule in the asymmetric unit including residues Q1 to L207, with clear electron density for residues A2 to A205 (Figure 3.2I, J and 3.14C). Like other cadherins, the structure of *hs* CDHR5 EC1-2 shows straight consecutive EC repeats, each with a typical seven β-strand Greek-key fold (Figure 3.2I, J). The linker region between the repeats has the canonical acidic residues that coordinate calcium and three calcium ions bound to it (Figure 3.2K and 3.19). Similar to PCDH15, a member of the Cr-3 subfamily of non-clustered protocadherins, *hs* CDHR5 EC1 has a disulfide bridge between cysteine residues C5 and C73 on strands A and F respectively (Figure 3.2L, 3.11C, and 3.14E). This disulfide bond keeps the N-terminal strand tucked to the protomer in a closed conformation, reminiscent of how calcium at site 0 facilitates closing of the N-terminal strand in EC1 repeats of CDH23 and *hs* PCDH24.

Next, we analyzed crystallographic contacts in the *hs* CDHR5 EC1-2 structure (Figure 3.20), which could lead to an understanding of how mouse, but not human CDHR5 mediates homophilic adhesion and how the heterophilic complex with PCDH24 is formed. The two largest contacts within the *hs* CDHR5 EC1-2 crystal structure are interfaces in antiparallel arrangements. The largest, at 1433.9 Å^2, is formed by interactions between the EC1 repeat of one protomer and EC2 of the adjacent one, with a slight shift in the register of the antiparallel dimeric interface (Figure 3.20A). There are up to three predicted N-glycosylation sites at this interface in the sequences for the human (p.N19 and p.N56) and the mouse (p.N17, p.N56, and p.N107) proteins. The next largest contact involves interactions between antiparallel EC2 repeats facing each other (Figure 3.20B), with an interface area of 530.2 Å^2, small relative to likely physiologically relevant interfaces (>856 Å^2), but comparable to interface area per EC repeat reported for physiologically relevant contacts in protocadherins [22,31,149]. In this arrangement, the EC3 repeat (not present in the structure), would interact with EC1, thus increasing the total contact area. None of the predicted N-glycosylation sites for the human or mouse sequences are at the EC2 interface of this *trans* dimer. An X-shaped crystallographic contact is also present (Figure 3.20C), with a small interface area (364.7 Å^2) involving EC1 to EC1 interactions. Additional smaller interfaces are present but are unlikely to be physiologically relevant given their atypical arrangement and small interface areas of 347.9 Å², 132.7 Å², and 114.0 Å², respectively (Figure 3.20D-F).

Although, *hs* CDHR5 does not mediate adhesion in bead aggregation assays, some of the contacts discussed above can serve as templates to predict the interfaces that could be used by *mm* CDHR5 to mediate homophilic adhesion. The largest, antiparallel EC1-2 interface might be impaired by glycosylation at a site predicted to exist in the mouse sequence (N107), but not in the human protein (R107). In addition, bead aggregation assays suggest that repeat EC3 is needed for adhesive interactions, suggesting that this EC1-2 interface is not driving CDHR5 homophilic adhesion. In contrast, the antiparallel EC1-3 interface mediated by EC2 interactions is compatible with results from bead aggregation assays indicating that the first three N-terminal repeats are sufficient for *trans* adhesiveness, but details of the arrangement are likely to be different due to the absence of EC3 in our structure and the low sequence conservation of surface residues (Figure 3.21). These contacts may also guide the search for interaction modes that facilitate the formation of PCDH24 and CDHR5 heterophilic complexes.

Human and mouse heterophilic PCDH24 and CDHR5 bonds are distinct

Previous bead aggregation assays showed that full-length extracellular domains of human PCDH24 and CDHR5 mediate heterophilic adhesion [67]. We used our library of Fc-tagged PCDH24 and CDHR5 cadherin fragments to test for heterophilic adhesiveness for both human and mouse proteins. To distinguish PCDH24 from CDHR5 coated beads, we used green and red fluorescent beads, respectively. We confirmed that human PCDH24 and CDHR5 cadherin extracellular domains mediate calcium-dependent heterophilic adhesion, and found that the mouse proteins also facilitate calcium-dependent bead aggregation (Figure 3.6A-B, I, J, K and 3.24A-B), suggesting that the heterophilic intermicrovillar cadherin bond driving intestinal brush border assembly is species independent.

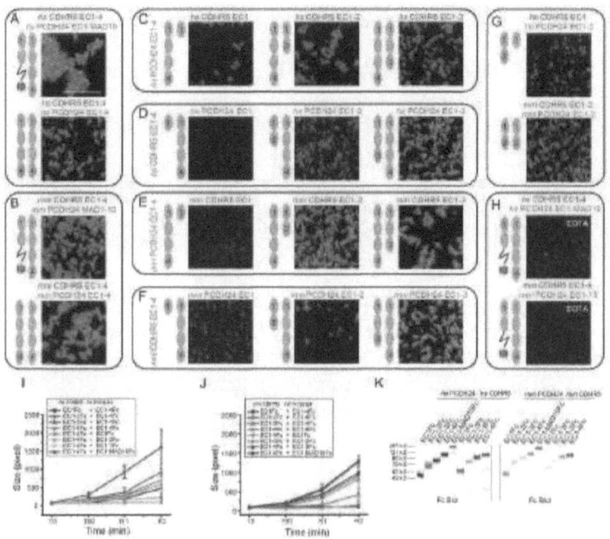

Figure 3.6 Heterophilic binding assays of hs and mm PCDH24 and CDHR5.
(A-B) Images from binding assays of hs PCDH24 (full-length extracellular domain and EC1-4Fc) mixed with the full-length hs CDHR5 cadherin extracellular domain (A), and mm PCDH24 (full-length extracellular domain and EC1-4Fc) mixed with full-length mm CDHR5 cadherin extracellular domain (B). Green fluorescent protein A beads are coated with PCDH24 fragments and red fluorescent protein A beads are coated with CDHR5 in all panels. All images are details of complete views shown in Figure 3.24. Images show bead aggregation observed after 60 min followed by rocking for 2 min in the presence of 2 mM CaCl$_2$. (C-F) Images from binding assays of truncations of hs CDHR5 mixed with hs PCDH24 EC1-4Fc (C), truncations of hs PCDH24 mixed with hs CDHR5 EC1-4Fc (D), truncations of mm CDHR5 mixed with mm PCDH24 EC1-4Fc (E), and truncations of mm PCDH24 mixed with mm CDHR5 EC1-4Fc (F). Images show bead aggregation observed after 60 min followed by rocking for 2 min in the presence of 2 mM CaCl$_2$. (G) Images from binding assays of minimum EC repeats required for heterophilic adhesion of hs and mm PCDH24 and CDHR5. The minimum units for heterophilic adhesion for the human proteins are CDHR5 EC1Fc and PCDH24 EC1-2Fc. The minimum units for heterophilic adhesion for the mouse proteins are CDHR5 EC1-2Fc and PCDH24 EC1-2Fc. Images show bead aggregation observed after 60 minutes followed by rocking for 2 minutes in the presence of 2 mM CaCl$_2$. (H) Protein A beads coated with full-length hs and mm PCDH24 and CDHR5 cadherin extracellular domains (including MAD10 for PCDH24 and without the mucin-like domain for CDHR5) in the presence of 2 mM EDTA shown as in (A-B). Bar - 500 μm. (I) Aggregate size for full-length and truncated constructs of hs PCDH24 and hs CDHR5 cadherin extracellular domains (including MAD10 for PCDH24 when indicated and without the mucin-like domain for CDHR5) at the start of the experiment (T0), after 60 min (T60) followed by rocking for 1 min (R1) and 2 min (R2). (J) Aggregate size for full-length and truncated constructs of mm PCDH24 and mm CDHR5 as in (I). Error bars in I and J are standard error of the mean (n indicated in Table 3.8). (K) Western blot shows expression and secretion of full-length and truncated versions of hs and mm PCDH24 and CDHR5 cadherin extracellular domains (including MAD10 for PCDH24 when indicated and without the mucin-like domain for CDHR5).

Next, we used truncated protein fragments and found that repeats EC1-4Fc of PCDH24 and CDHR5 from both species are sufficient to drive heterophilic bead aggregation (Figure 3.6A-B, I, J, K and 3.24A-B). Further bead aggregation assays focused on determining the minimum adhesive units for heterophilic adhesion. We carried out heterophilic binding assays using PCDH24 EC1-4Fc mixed with truncated fragments of CDHR5 and vice versa for both human and mouse proteins (Figure 3.6C-F, I, J, K and 3.24C-F). The minimal adhesive units for heterophilic binding when using human proteins were PCDH24 EC1-2Fc and CDHR5 EC1Fc, whereas mouse proteins required PCDH24 EC1-2Fc and CDHR5 EC1-2Fc.

Additional experiments using only the required EC repeats of PCDH24 and CDHR5 to test heterophilic adhesiveness for both human and mouse proteins confirmed the results obtained with the truncation series (Figure 3.6G, K and 3.24G, I). Binding assays in the presence of 2 mM EDTA abolished aggregation indicating that heterophilic adhesion mediated by PCDH24 and CDHR5 is calcium-dependent (Figure 3.6H, K and 3.24H, J). The low sequence similarity of CDHR5 and PCDH24 across different species may explain the differences in their binding mechanisms, as demonstrated by the distinct minimal adhesive units found for the human and mouse heterophilic complexes. Remarkably, while homophilic adhesive behavior is not conserved between human and mouse, the heterophilic bond is robust with slight differences in underlying molecular mechanisms.

Discussion

Our sequence analyses, structures, and bead aggregation assays suggest that there are various modes of *trans* homophilic and heterophilic interactions involving the extracellular domains of PCDH24 and CDHR5 (Figure 3.7). We have confirmed that human PCDH24 can form homophilic bonds [67], and found three potentially distinct modes of *trans* adhesion in bead aggregations assays. The first mode involves weak (small bead aggregates) calcium-independent interactions mediated by EC1. This adhesive mode is not strong enough to drive calcium-dependent PCDH24 adhesion when using larger parts of its extracellular domain, and therefore might be unphysiological or relevant only when a pre-formed bond is exposed to calcium chelators. The physiological PCDH24 *trans* homophilic bond might be initiated by weak interactions involving EC1 and EC2 repeats (small bead aggregates), with further interdigitation for stronger interactions (larger bead aggregates) mediated by at least EC1 to EC3 repeats.

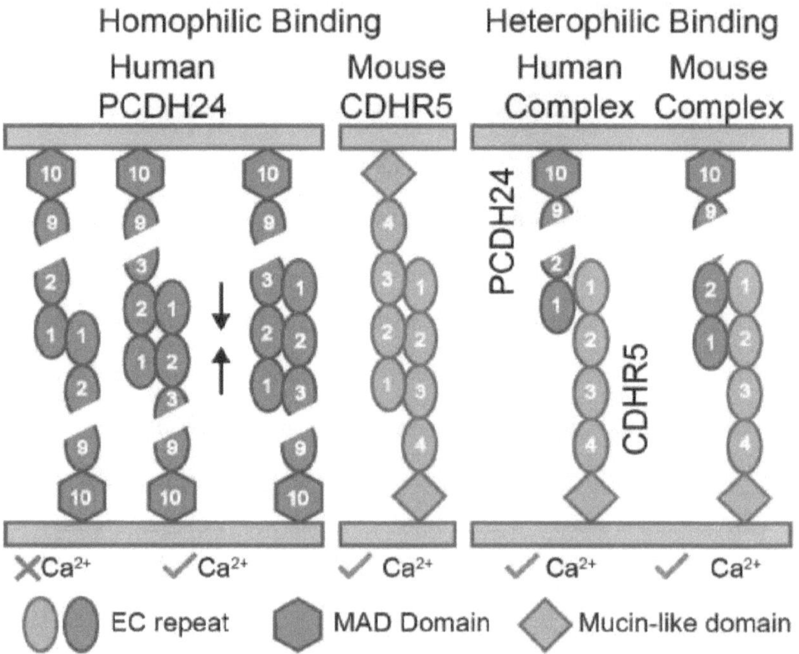

Figure 3.7 Schematics of homophilic and heterophilic adhesion mediated by human and mouse PCDH24 and CDHR5.
Homophilic adhesion that results in small bead aggregates can be mediated by *hs* PCDH24 EC1 (calcium independent) and EC1-2 (calcium dependent), while homophilic adhesion that results in large bead aggregates requires *hs* PCDH24 EC1-3 (calcium dependent). Homophilic adhesion mediated by *mm* CDHR5 requires EC1-3 (calcium dependent). Heterophilic adhesion is mediated by PCDH24 EC1-2 and CDHR5 EC1 when using human proteins and by PCDH24 EC1-2 and CDHR5 EC1-2 when using mouse proteins (both calcium dependent).

Intriguingly, our bead aggregation assays indicate that mouse PCDH24 does not mediate *trans* homophilic adhesion, highlighting a species-dependent behavior that is consistent with poor sequence and structural conservation for the mouse and human proteins. In contrast, experiments with CDHR5 reveal that the mouse protein can mediate *trans* homophilic adhesion, but the human protein cannot. These results are consistent with L929 cell aggregation assays that used the rat CDHR5 protein [61,196]. The physiological relevance of the *trans* homophilic bonds for human PCDH24 and mouse CDHR5 remains to be determined. Mice lacking PCDH24 expression in the intestinal track are viable, but intestinal tissue exhibits various structural defects leading to functional impairment of intestine function and loss of body weight [103]. Homophilic CDHR5 bonds might be partially compensating for the lack of heterophilic intermicrovillar links in these mice.

Importantly, adhesion mediated by PCDH24 and CDHR5 heterophilic bonds is species independent in our bead aggregation assays with mouse and human proteins. These results also demonstrate that lack of homophilic adhesion for mouse PCDH24 and for human CDHR5 is not caused by issues with bead coating or with protein misfolding and degradation in our assays. Details of the heterophilic interaction between PCDH24 and CDHR5, however, might differ from species to species, as we found that the minimal adhesive units are PCDH24 EC1-2Fc / CDHR5 EC1Fc for human proteins and PCDH24 EC1-2Fc / CDHR5 EC1-2Fc for mouse proteins.

Our structures of the human PCDH24 EC1-2, mouse PCDH24 EC1-3, and human CDHR5 EC1-2 protein fragments confirm the expected structural similarities between the PCDH24 and CDH23 tips, and between the PCDH15 and CDHR5 tips. These structures also support binding mechanisms that are distinct from those used by classical cadherins, clustered protocadherins, and non-clustered δ protocadherins [8,22,27,29,31,32,51,194,195]. While crystal contacts in our structure suggest possible interfaces that might facilitate the modes of adhesion discovered through bead aggregation assays, further work will be required to elucidate and validate structural models of *trans* homophilic and heterophilic complexes, to understand the relevance of an *in crystallo cis* complex observed here for mouse PCDH24, to clarify the structural role played by glycosylation *in vivo* [197], and to determine which complexes form intermicrovillar links *in vivo*.

Intermicrovillar links observed in human CACO-2$_{BBE}$ cells and in native mouse tissue are 46.8 ± 8.9 nm and 49.9 ± 8.8 nm in length, respectively [67]. Interestingly, lengths range from 30 to 80 nm for both human and mouse links, perhaps reflecting the challenges of measuring link sizes from two-dimensional electron-microscopy images of cells, with some additional uncertainty coming from difficult-to-define link edges. Nevertheless, this range of lengths hints at multiple possibilities for the architecture of the complex that are consistent with the modes of adhesion we found using bead aggregation assays. Links made of two PCDH24 molecules interacting in *trans* and tip-to-tip could reach lengths of up to ~90 nm (9 EC repeats + MAD10 per molecule; 4.5 nm long each), while links made of the shortest CDHR5 isoforms with just 4 EC repeats interacting tip-to-tip could be 36 nm in length or even shorter if there is any overlap at the junction or bending at any of the EC linkers. Heterophilic links formed by a tip-to-tip *trans* interaction of the full-length PCDH24 ectodomain and the shortest CDHR5 isoform should be ~63 nm or shorter. The average value observed *ex vivo* and *in vivo* (~48 nm) indicates that some overlap at the junction may exist, and that most of the links are likely formed by the heterophilic complex. The minimal adhesive units we found for mouse and human PCDH24 and CDHR5 homophilic and heterophilic *trans* complexes are consistent with these estimates for length ranges, assuming that up to 3 EC repeats might be overlapping to form robust bonds that lead to bead aggregation *in vitro*, and adhesion *in vivo*.

The functional role of PCDH24 and CDHR5 in non-intestinal tissues and cells is unclear. These proteins have also been found in the liver and kidney [196,198] and have been implicated in contact inhibition of cell proliferation [198,199], morphogenesis [192,196,197], colorectal carcinogenesis [200,201], and gallstone disease [202]. Exploring how different organs from different species express and use PCDH24 and CDHR5 proteins, which have poorly conserved sequences, might provide with a unique opportunity to reveal how evolutionary events structurally encoded adhesive and signaling functions in various physiological contexts [203].

Materials and Methods

Cloning, expression, and purification of PCDH24 and CDHR5 fragments for crystallography

DNA encoding for *hs* PCDH24 EC1-2, *mm* PCDH24 EC1-3 and *hs* CDHR5 EC1-2 were sub cloned into *NdeI* and *XhoI* sites of the pET21a vector including a fused C-terminal hexa-histidine tag. All DNA constructs were sequence verified. These constructs were used to transform BL21(DE3) pLysS cells (Stratagene). The PCDH24 transformants were cultured in TB and induced at $OD_{600} \sim 0.6$ with 1 mM IPTG at 37°C overnight. The *hs* CDHR5 EC1-2 transformant was cultured in TB and induced at an $OD_{600} \sim 0.6$ with 1 mM IPTG at 30°C overnight. Cells were lysed by sonication in denaturing buffer (20 mM TrisHCl [pH7.5], 6 M guanidine hydrochloride [GuHCl], 10 mM $CaCl_2$ and 20 mM imidazole). The cleared lysates were loaded onto Ni-Sepharose beads and nutated for ~1 h (GE Healthcare). Beads were washed twice with denaturing buffer and hexa-histidine-tagged proteins were eluted with the same buffer supplemented with 500 mM imidazole. After purification, 2 mM DTT was added to the protein before refolding. The *hs* PCDH24 EC1-2 protein was refolded in six dialysis steps using a buffer with 20 mM TrisHCl [pH 8.0], 5 mM $CaCl_2$ and decreasing amounts of GuHCl (6 M, 3 M, 2 M, 1 M, 0.5 M and 0 M). In steps four, five and six the refolding buffer included 400 mM arginine. Steps three through six also included 150 mM NaCl. The *mm* PCDH24 EC1-3 and *hs* CDHR5 EC1-2 proteins were refolded overnight using a buffer with 400 mM Arginine, 20 mM TrisHCl [pH 8.0], 150 mM NaCl, and 5 mM $CaCl_2$. All dialysis steps were performed using MWCO 2000 membranes and used refolding buffers for 12 hours at 4°C. Refolded protein was further purified on a Superdex200 16/600 column (GE Healthcare) in 20 mM TrisHCl [pH

8.0], 150 mM NaCl and 2 mM CaCl$_2$, and concentrated by ultrafiltration to ~5 mg/ml for crystallization (Vivaspin 10 KDa).

Crystallization, data collection, and structure determination

Crystals were grown by vapor diffusion at 4°C by mixing 1 µL of protein and 0.5 µL of reservoir solution including (0.2 M MgCl, 0.1 M HEPES [pH 7.3], 10% PEG 4000) for *hs* PCDH24 EC1-2, (3.0 M LiCl, 0.1 M HEPES [pH 7.5]) for *mm* PCDH24 EC1-3 and (0.1 M Na Acetate pH 4.4, 0.7 M CaCl$_2$) for *hs* CDHR5 EC1-2. Crystals were cryo-protected in reservoir solution plus a cryoprotectant (25% PEG 400 for PCDH24 and 25% glycerol for *hs* CDHR5 EC1-2) and cryo-cooled in liquid N$_2$. X-ray diffraction datasets were collected as indicated in Table 3.1 and processed with HKL2000 [160].

All structures were solved by molecular replacement using PHASER [161]. The first two N-terminal repeats of CDH23 EC1-2 (PDB: 4AQE) [33] were used as an initial model for the *mm* PCDH24 EC1-3 structure. The *mm* PCDH24 EC3 repeat was built using BUCCANEER [204,205]. The *hs* PCDH24 EC1-2 structure was determined by using the first two N-terminal EC repeats of the *mm* PCDH24 EC1-3 structure (PDB: 5CYX). For the *hs* CDHR5 EC1-2 structure we used a model of CDH23 EC22-24 (PDB: 5UZ8) [24] created using RaptorX [206]. Model building was done with COOT [207] and refinement was performed with REFMAC5 [159]. Restrained TLS refinement was used for the PCDH24 structures. Data collection and refinement statistics are provided in Table 3.1.

Cloning of Fc-fusions proteins for bead aggregation assays

DNA constructs encoding C-terminal truncations of *hs* and *mm* PCDH24 as well as of *hs* and *mm* CDHR5, all fused to a C-terminal Fc-tag, were prepared for bead aggregation assays. Truncations were made at DXNDN or similar ends of each EC repeat (Figure 3.8 and 3.9). Truncations were cloned into a modified CMV vector (Jontes laboratory, OSU) using *XhoI* and *KpnI* (*hs* PCDH24, *mm* PCDH24), *NheI* and *XhoI* (*hs* CDHR5), and *XhoI* and *BamHI* (*mm* CDHR5) cutsites. The sequence for the linker between the *hs* CDHR5 protein fragments and their Fc tags is LELKLRILQSTVPRARDPPV, while the sequence for the linker between the *mm* CDHR5 protein fragments and their Fc tags is TDPPV. Mutations were generated using the QuikChange Lightning kit (Agilent). All DNA constructs were sequence verified.

Bead aggregation assays

Bead aggregation assays were performed as previously described [22,31,45,67,162]. The PCDH24Fc and CDHR5Fc fusion constructs were transfected into HEK293T cells via calcium-phosphate transfection [164,165,208]. A solution of 10 μg of DNA and 250 mM $CaCl_2$ was added dropwise to 2X HBS with mild vortexing. The solution was immediately added drop-wise to a 100 mm dish of HEK293T cells cultured in DMEM with FBS, L-glut, and PenStrep. For each construct, two plates of cells were transfected. The following day, cells were rinsed twice with 1X PBS and allowed to grow in serum-free media for two days prior to collection. The collected media contained the secreted Fc fusion proteins and was concentrated (Amicon 10 kD, 30 kD concentrators) to a volume of 500 μL before being incubated with 1.5 µL of Protein G Dynabeads (Invitrogen). The beads were incubated with rotation for 2 hr at 4°C. Incubated beads were washed with binding buffer (50 mM Tris [pH 7.5], 100 mM NaCl, 10 mM KCl, and 0.2% BSA) and split into two tubes, either with 2 mM $CaCl_2$ or 2 mM EDTA. Beads were then allowed to aggregate on a glass depression slide in a humidified chamber for 1 h, then rocked for 1 to 2 min at 8 oscillations/min. Images were taken at various time points using a Nikon Eclipse Ti microscope with a 10X objective as detailed in Table 3.8. Sizes of bead aggregates were quantified with ImageJ as described previously [45,162]. Images were thresholded to exclude background and the area of the detected particles was measured in pixels. Average aggregate size (± SEM) was plotted in Origin Pro. Number of independent biological replicates are listed in Table 3.8.

Fluorescent bead aggregation assays

Fluorescently labeled protein A Dynabeads were produced for the heterophilic bead aggregation assays. Protein A (5 mg; Thermo Fisher Pierce) was resuspended in 1X PBS to obtain a final concentration of 2 mg/mL. Using the Thermo Fisher Dylight Antibody Labeling kits, 1 mg of Protein A was labeled with Dylight 488 to produce 0.5 moles of dye/1 mole of Protein A. Labeled Protein A (200 µg) was then conjugated to Dynabeads M-280 Tosylactivated (Invitrogen) following the provided protocol to produce 10 mg of beads at a final concentration of 20 mg/mL. A similar protocol was used to obtain Dylight 594 conjugated Protein A Dynabeads. Conjugation to the beads was confirmed by observing the beads in depression slides under 10X magnification with the GFP filter cube

(Nikon) for the Dylight 488 fluorophore and the Texas Red filter cube (Nikon) for the Dylight 594 fluorophore. Beads of both colors were then mixed on a depression slide and no overlap of fluorescence was observed. Further confirmation that the beads were functional was done by performing a bead aggregation assay with a positive control (*hs* PCDH24 EC1-4).

Fluorescent bead aggregation assays were performed as done with non-fluorescent beads but with noted exceptions. Concentrated media was added to 2.2 μL of fluorescently labeled beads to achieve the same concentration as the Protein G Dynabeads on the slides. Dylight 488 conjugated beads were used for all PCDH24Fc fusions and Dylight 594 conjugated beads were used for all CDHR5Fc fusions. Prior to addition of beads to the CaCl$_2$ or EDTA tubes, 160 μL of Dylight 488 beads containing the PCDH24 constructs were mixed with 160 μL Dylight 594 beads containing CDHR5 constructs, then 150 μL of the mixture was added to the CaCl$_2$ or EDTA tubes.

Western Blots

Western blots were performed for Fc-tagged proteins to confirm expression and secretion of the protein. Protein samples were mixed with 4X SDS with β-mercaptoethanol, boiled for 5 min, and loaded onto SDS-PAGE gels (BioRad) for electrophoresis. Proteins were electroblotted onto PVDF membrane (GE Healthcare), blocked with 5% nonfat milk in TBS with 0.1% Tween-20 for 1 h, and incubated overnight at 4°C with goat anti-human IgG (1:200; Jackson ImmunoResearch Laboratories, Catalog-109-025-003, Lot-117103,

135223). After washing with 1X TBS, the membranes were incubated with mouse anti-goat HRP-conjugated secondary antibody (1:5000; Santa Cruz Biotechnology, Catalog-sc-2354, Lot-A2017, J1718) for 1 h at room temperature. Membranes were washed in 1X TBS and developed using the ECL Select Western Blot Detection kit (GE Healthcare) for chemiluminescent detection (Omega Lum G). Western blots were performed on samples obtained immediately after media collection and concentration as well as protein samples obtained from the Protein G/A beads following aggregation assays.

Sequence alignments

Sequence alignments were performed on individual EC repeats for a variety of species of PCDH24 (Table 3.3) and CDHR5 (Table 3.4) in MUSCLE [167]. Sequences that covered the complete extracellular domain were selected to include a variety of species of each protein. Sequence conservation per EC repeat in percent identity was obtained from the MUSCLE alignments in Geneious [209] (Table 3.2). All EC repeats of both human and mouse PCDH24 (Figure 3.12) and CDHR5 (Figure 3.19) were aligned to one another as well as described above. Conservation of alignments were visualized in JalView [168] using percent identity [210] with a conservation cutoff of 40%. Additionally, sequence alignments of individual EC repeats for a variety of species of PCDH15 [36] (Table 3.5) and CDH23 [24] (Table 3.6) were performed in MUSCLE, and sequence conservation per EC repeat was obtained in Geneious (Table 3.2). A separate set of alignments, which include up to 5 species for the N-terminal repeats (EC1-3) of PCDH24, CDHR5, CDH23, PCDH15, CDH1, and CDH2 (Table 3.7), were performed in MUSCLE. The Sequence Identity and Similarity Server

(SIAS) was used to determine identity between pairs of different species for each protein (Figure 3.1B). A broader conservation analysis of PCDH24 EC1-3 and CDHR5 EC1-2 was done in Consurf [170–174] using additional sequences to include a larger variety of species (Table 3.9 for PCDH24 and Table 3.10 for CDHR5). The values obtained from Consurf (1 = lowest conservation, 9 = highest conservation) were matched to the corresponding residues in the structures and used to make a heat map of conservation. N-glycosylation and O-glycosylation sites for the extracellular domains of PCDH24 and CDHR5 were predicted using NetNGlyc 1.0 and NetOgly 4.0 [211] and highlighted in the multiple sequence alignment. All positives for O-glycosylation were labeled, but predictions for N-glycosylation were only labeled if the prediction received a 9/9 prediction score from the server.

Simulated systems

The psfgen, solvate, and autoionize VMD [177] plugins were used to build all systems (Table 3.11), adding hydrogen atoms to the protein structure and crystallographic water molecules. For the *mm* PCDH24 EC1-3 systems, the missing loop in EC1 (residues 32 to 37) was built in COOT [207] using the human PCDH24 EC1 structure as a template. Non-native N-terminal residues were removed. Residues D, E, K, and R were considered to be charged and neutral histidine protonation states were determined on favorable hydrogen bond formation. The proteins were placed in additional water and randomly placed ions to solvate and ionize the systems at 150 mM NaCl. For SMD simulations, protein molecules were aligned to the x-axis. Additional information for systems is presented in (Table 3.11).

Simulations Parameters

Equilibrium and SMD simulations were performed using NAMD 2.11 [212], the CHARMM36 forcefield for proteins with the CMAP correction, and the TIP3P model for water [213]. Van der Waals interactions were computed with a cutoff at 12 Å and using a switching function starting at 10 Å. Computation of long-range electrostatic forces was done with the Particle Mesh Ewald method with a cutoff of 12 Å and a grid point density of >1 Å^{-3}. A 2 fs uniform integration time was used with SHAKE. When indicated, constant temperature ($T = 300K$) was enforced using Langevin dynamics with a dampening coefficient of 0.1 ps^{-1} unless otherwise noted. The hybrid Nose-Hoover Langevin piston method was used to maintain constant pressure (NpT) at 1 atm with a 200 fs decay period and a 100 fs dampening time constant. In simulation S3a (Table 3.11), the *mm* PCDH24 EC1-3 protein had harmonic constraints ($k = 1$ kcal mol^{-1} Å^{-2}) applied to the C$_\alpha$ atoms of residues S110, V123, and L182 in EC2 to avoid rotations leading to interactions between periodic images. Constant velocity stretching simulations (S1b-d and S3b-d, Table 3.11) were carried out using the SMD method and the NAMD Tcl interface [214–217]. Virtual springs ($k_s = 1$ kcal mol^{-1} Å^{-2}) were attached to Cα atoms of C-terminal ends for unbinding simulations (S1b-d) and to the center of mass of EC1 and EC3 for unfolding simulations (S3b-d). The free ends of the stretching springs were moved in opposite directions, away from the protein(s), at a constant velocity. Applied forces were computed using the extension of the virtual springs. Maximum force peaks and their averages were computed from 50 ps running averages.

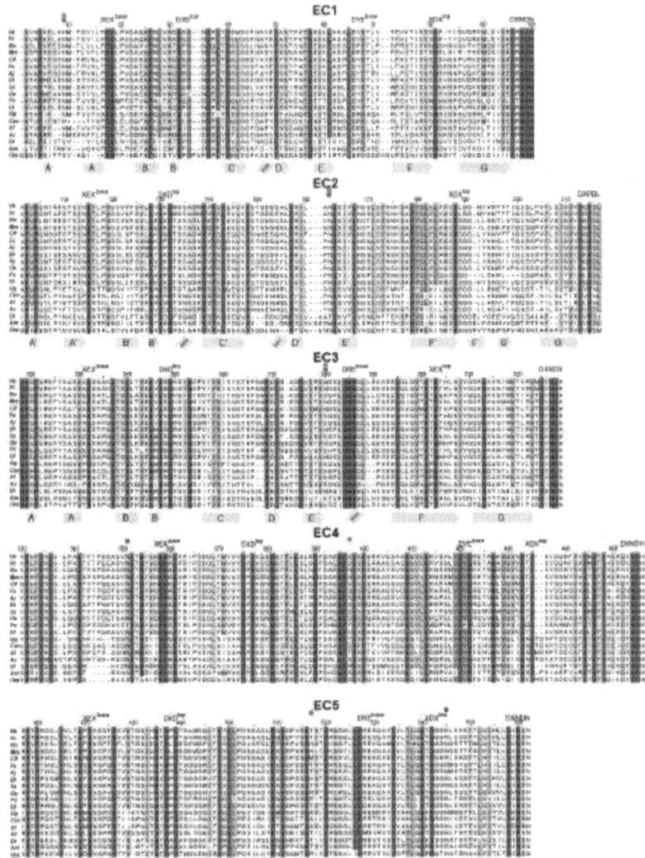

Figure 3.8 Sequence alignments of individual EC repeats of PCDH24.
Multiple sequence alignments comparing each EC repeat and MAD10 of PCDH24 from 19 different species. Each alignment is colored by percent identity, with white being the lowest percent identity and dark blue being the highest. Sites of N-linked (blue in human, green in mouse) and O-linked (purple in human, red in mouse) glycosylation are denoted by a colored circle. Secondary structure elements observed in the crystal structures of *hs* PCDH24 EC1-2 and *mm* PCDH24 EC1-3 are illustrated below the respective repeats. Calcium-binding motifs are indicated above the sequences, which are numbered according to the human protein. Species are abbreviated as follows: *Homo sapiens (hs), Pan troglodytes (Pt), Rattus norvegicus (Rn), Mus musculus (mm), Canis lupus familiaris (Clf), Felis catus (Fc), Acinoyx jubatus (Aj), Bos Taurus (Bt), Sus scrofa (Ss), Ovis aries (Oa), Phascolarctos cinereus (Pc), Delphinapterus leucas (Dl), Gallus gallus (Gg), Corvus cornix cornix (Ccc), Aptenodytes forsteri (Af), Anolis carolinensis (Ac), Danio rerio (Dr), Astyanax mexicanus (Am),* and *Oncorhynchus mykiss (Om).* Species were chosen based on sequence availability and taxonomical diversity. Accession numbers and species can be found in Table 3.3.

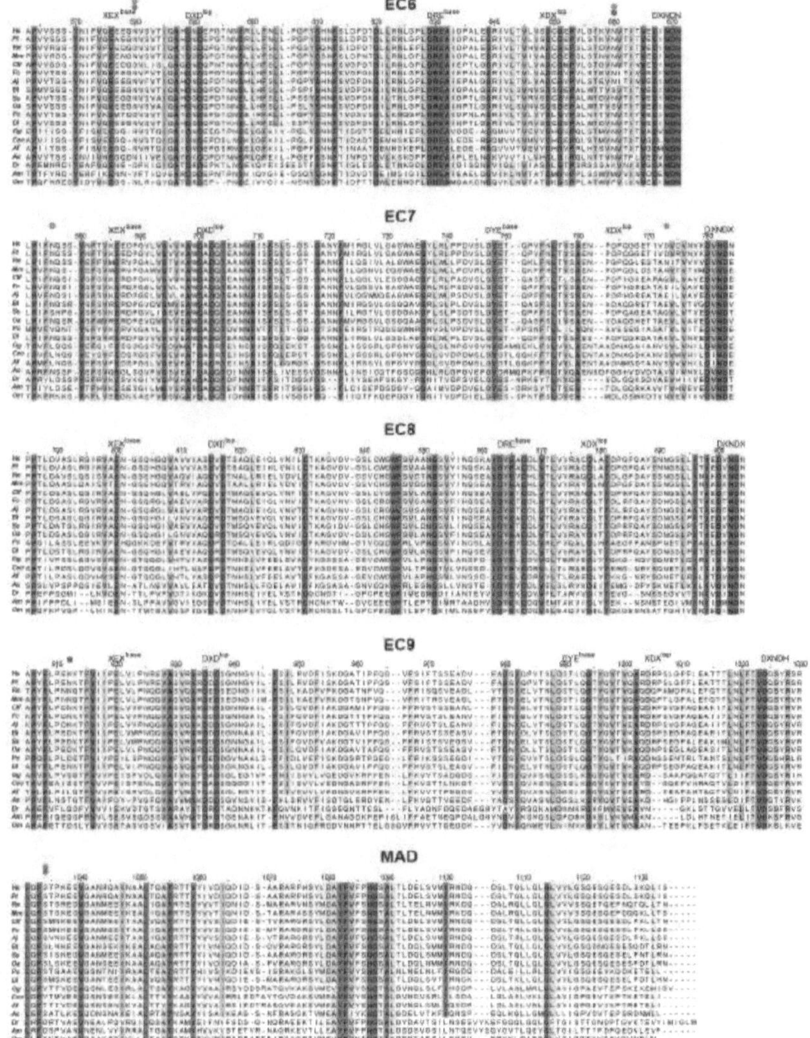

EC6
EC7
EC8
EC9
MAD

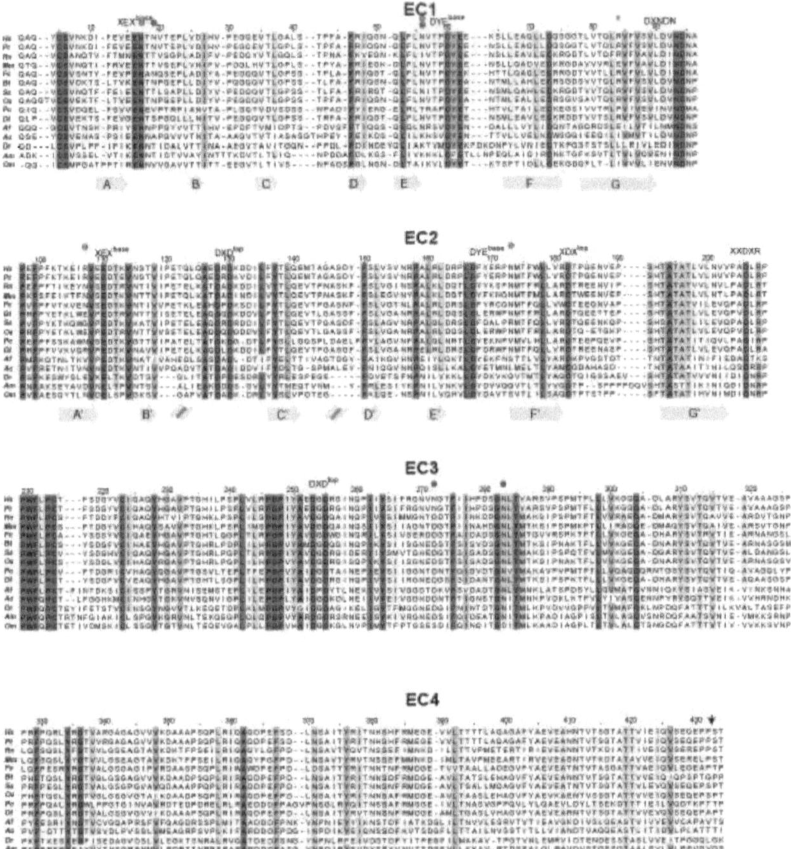

Figure 3.9 Sequence alignments of individual EC repeats of CDHR5.

Multiple sequence alignments comparing each EC repeat of CDHR5 from 15 different species, shown as in Figure 3.8. An asterisk (*) indicates site R84 mutated in binding assays. An arrow indicates the end of EC1-EC4 protein fragments used in binding assays. Secondary structure elements observed in the crystal structures of *hs* CDHR5 EC1-2 are illustrated below the respective repeats. Calcium-binding motifs are indicated above the sequences, which are numbered according to the human protein. Species are abbreviated as follows: *Homo sapiens (hs), Pan troglodytes (Pt), Rattus norvegicus (Rn), Mus musculus (mm), Felis catus (Fc), Bos Taurus (Bt), Sus scrofa (Ss), Ovis aries (Oa), Phascolarctos cinereus (Pc), Delphinapterus leucas (Dl), Aptenodytes forsteri (Af), Anolis carolinensis (Ac), Danio rerio (Dr), Astyanax mexicanus (Am),* and *Oncorhynchus mykiss (Om).* Species were chosen based on sequence availability and taxonomical diversity. Accession numbers and species can be found in Table 3.4.

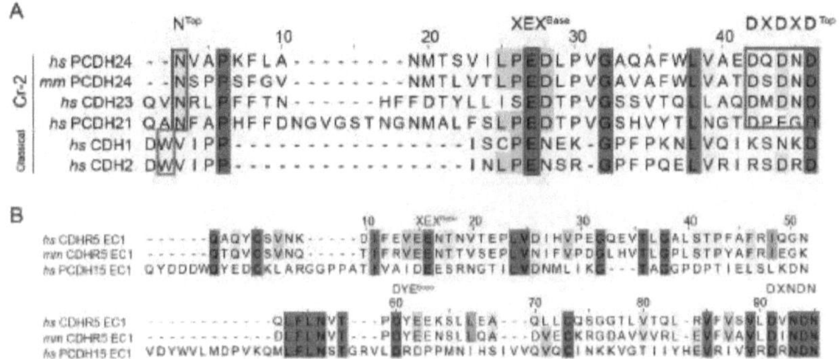

Figure 3.10 Comparison of PCDH24 and CDHR5 N-termini with other cadherins.
(A) Alignment of the processed N-terminal sequences of Cr-2 protocadherins *hs* PCDH24, *mm* PCDH24, *hs* CDH23 and *hs* PCDH21, and of two classical cadherins, *hs* CDH1 and *hs* CDH2. Calcium-binding sites are labeled and boxed in red in the sequence to show the additional calcium-bindings sites in Cr-2 protocadherins not present in classical cadherins. Residue W2 involved in classical cadherin binding is boxed in pink. (B) Alignments of the processed sequences of EC1 for *hs* CDHR5, *mm* CDHR5, and *hs* PCDH15. Calcium-binding sites are labeled. The two cysteine residues that form a disulfide bond between β-strands A and F are highlighted by a red box.

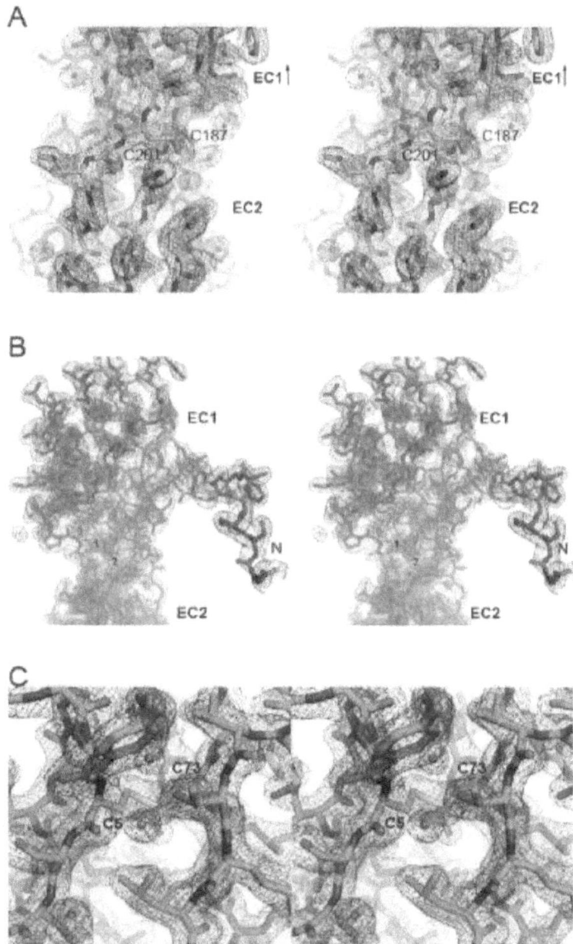

Figure 3.11 Electron density for unique elements of *hs* PCDH24 EC1-2, *mm* PCDH24 EC1-3 and *hs* CDHR5 EC1-2.
(A) Extended F-G loop of EC2 in *hs* PCDH24 where the disulfide bond is located (C187 - C201) shown at 1.0 σ using carve of 1.6. This loop is also present in the *mm* PCDH24 structure. (B) The atypical extended N-terminus seen in *mm* PCDH24 EC1 shown at 1.0 σ and carve of 1.8. (C) Disulfide bond between residue C5 in β-strand A and residue C73 in β-strand F of *hs* CDHR5 EC1-2 shown at 1.0 σ and carve of 1.6.

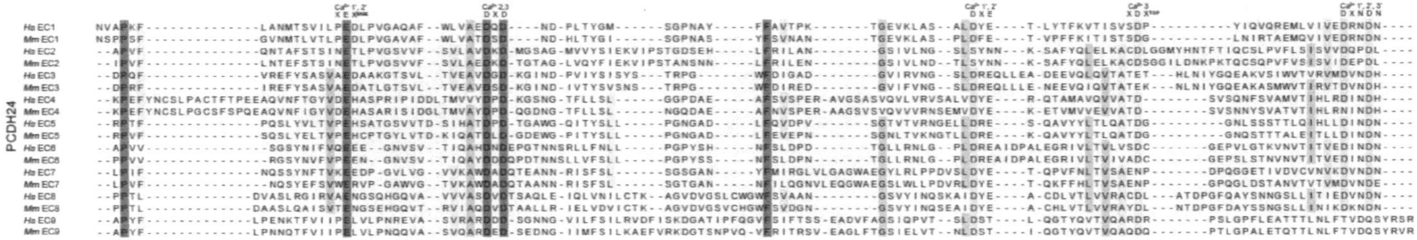

Figure 3.12 Sequence alignment of hs and mm PCDH24 EC repeats.
All 9 EC repeats for each species are aligned to each other (EC1-9). Conserved calcium-binding motifs are labeled on top of the alignment.

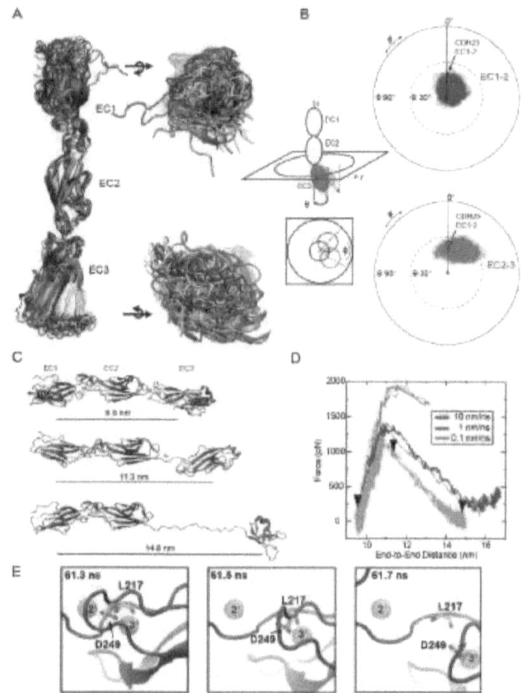

Figure 3.13 Equilibrium and stretching simulations of *mm* PCDH24 EC1-3.
(A) Superposition of *mm* PCDH24 EC1-3 conformations taken every 5 ns from a ~99 ns long trajectory (simulation S2, Table 3.11). Repeat EC2 was used as a reference. Side, top, and bottom views are shown. Color indicates time step (red-white-blue). (B) Orientation projections illustrating the conformational freedom of the EC1-2 and EC2-3 linkers throughout equilibrium simulations. To quantify the conformational freedom of EC2 relative to EC1 (top), and of EC3 relative to EC2 (bottom), the longest principal axes of EC1 and EC3 were aligned to the z axis and then the projections of the longest principal axes of EC2 (blue) and EC3 (red) in the x-y plane were plotted. The initial orientation of a control protomer, CDH23 EC1-2 (2WHV), is shown as a black dot [110]. The EC2-3 linker behavior is not dramatically different than the behavior observed for the EC1-2 linker, suggesting similar flexibility. (C) Trajectory snapshots during the slowest speed stretching simulation at 0.1 nm/ns for *mm* PCDH24 EC1-3 (S3d, Table 3.11). Springs indicate position (center of mass) and direction of applied forces. Unfolding of the EC2-3 linker is observed first. End-to-end distances between the centers of mass of EC1 and EC3 are indicated for each snapshot. (D) Force versus end-to-end distance for simulations of the *mm* PCDH24 EC1-3 monomer (S3a-d) at stretching speeds of 10 nm/ns (red), 1 nm/ns (blue) and 0.1 nm/ns (green). Dark and light colors indicate forces applied at opposite ends. Black arrowheads indicate time points illustrated in (C). (E) Detail of the *mm* PCDH24 EC2-3 linker at the force peak during the slowest stretching simulation at 0.1 nm/ns (S3d, Table 3.11). Residues involved in two-step unbinding from calcium ions are shown.

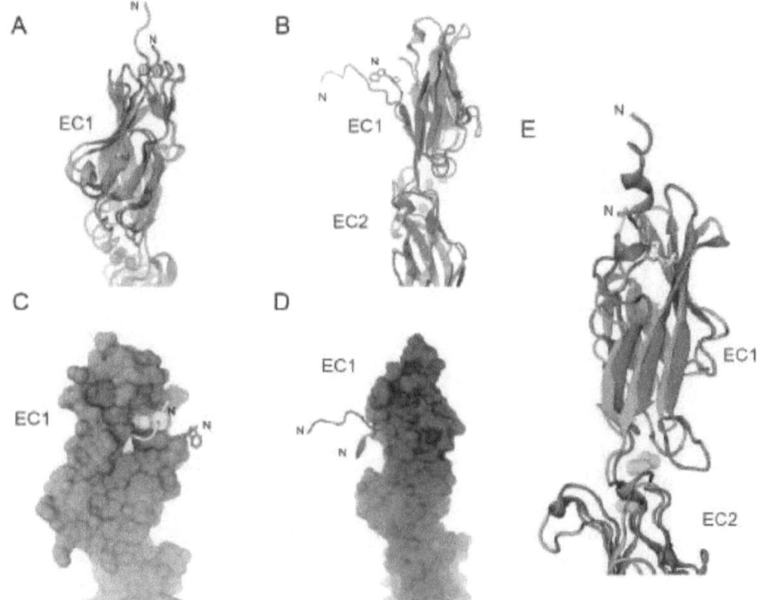

Figure 3.14 Structural comparison of PCDH24 and CDHR5 EC1 repeats with other cadherins.
(A) Detail of superposed EC1 N-termini of *hs* PCDH24 (blue) and *mm* CDH23 (grey; PDB: 2WHV). Both have a calcium ion at the calcium-binding site 0, a feature of Cr-2 protocadherins. (B) Detail of superposed EC1 N-termini of *mm* PCDH24 (blue) and *hs* CDH1 (grey; PDB: 2O72). Both have N-termini protruding away from the protomer to interact with another neighboring EC1, as shown in (C) for *hs* CDH1 and (D) for *mm* PCDH24. (C-D) Surface representations of *hs* CDH1 and *mm* PCDH24, respectively. Highlighted in yellow is β-strand A with W2 from its binding partner. Residues 25, 26, and 27 are not shown to facilitate visualization of the tryptophan binding pocket. Highlighted in purple is the N-terminus of *mm* PCDH24 from a neighboring protomer in the crystal lattice (Figure 3.2D). (E) Detail of *hs* CDHR5 EC1 (green) and *mm* PCDH15 EC1 (purple) superposed. Disulfide bonds are shown in orange and yellow. PCDH15 has longer loops and a longer N-terminus, with a helix that interacts with CDH23.

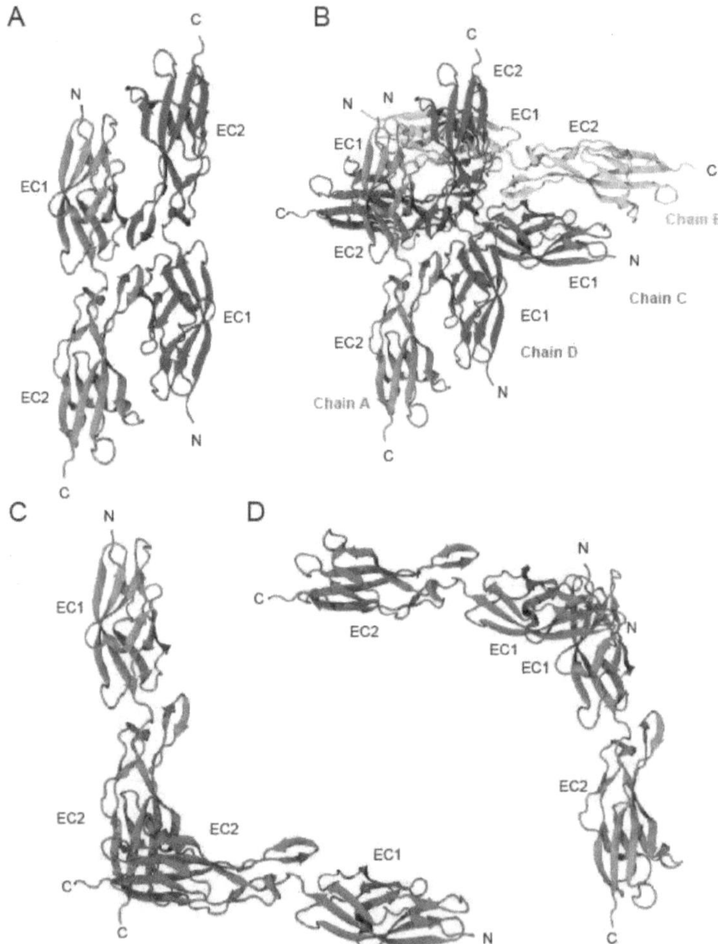

Figure 3.15 Crystal contacts in the *hs* PCDH24 EC1-2 structure.
(A-D) Ribbon diagram of two protomers showing a crystal contact between chains A and D with an interface area of 979.2 Å². The arrangement is antiparallel, likely describing a possible *trans* interface (A). Crystal contacts in the entire asymmetric unit include four additional interfaces with areas of 173.6 Å² (chains A and B), 231.1 Å² (chains A and C), 255.3 Å² (chains D and B), and 307.2 Å² (chains D and C). Two Additional interfaces between chains A and C (387.4 Å² and 507.3 Å²) are shown in (C) and (D). Equivalent interfaces for the one shown in (A) between chains B and C, and those shown in (C, D) between chains D and B, have similar interface areas of 938.6 Å², 390.8 Å², and 515.1 Å², respectively, but are not shown.

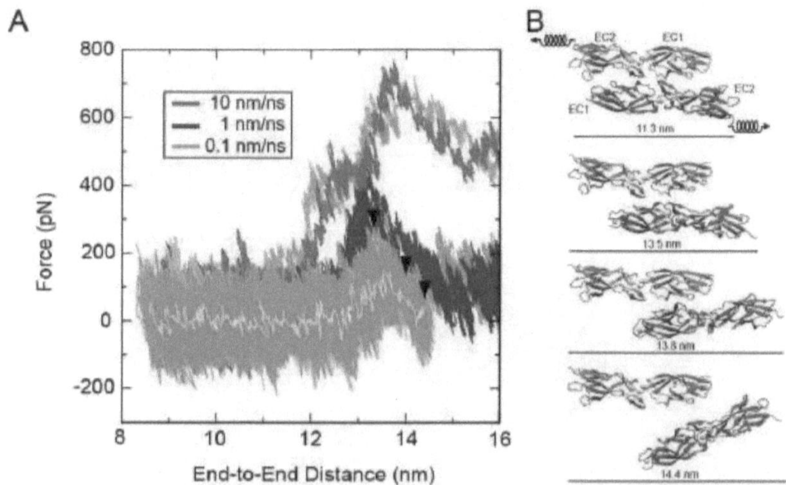

Figure 3.16 Stretching simulations testing the strength of the *hs* PCDH24 EC1-2 *trans* interface.
(A) Force versus end-to-end distance for simulations of the largest crystallographic *trans* interface (3.15A) observed for *hs* PCDH24 EC1-2 (simulations S1b-S1d, Table 3.11). Forced unbinding simulations were carried out at stretching speeds of 10 nm/ns (red), 1 nm/ns (blue), and 0.1 nm/ns (green, 50 ps running average in light green). (B) Snapshots of unbinding trajectory during stretching simulation at 0.1 nm/ns (S1d, Table 3.11). Springs indicate position and direction of applied forces. Top panel shows complex at the beginning of the simulation, other panels show snapshots at three time points indicated with black arrow heads in (A). End-to-end distance indicates the distance between the stretched atoms on the C-termini of EC2.

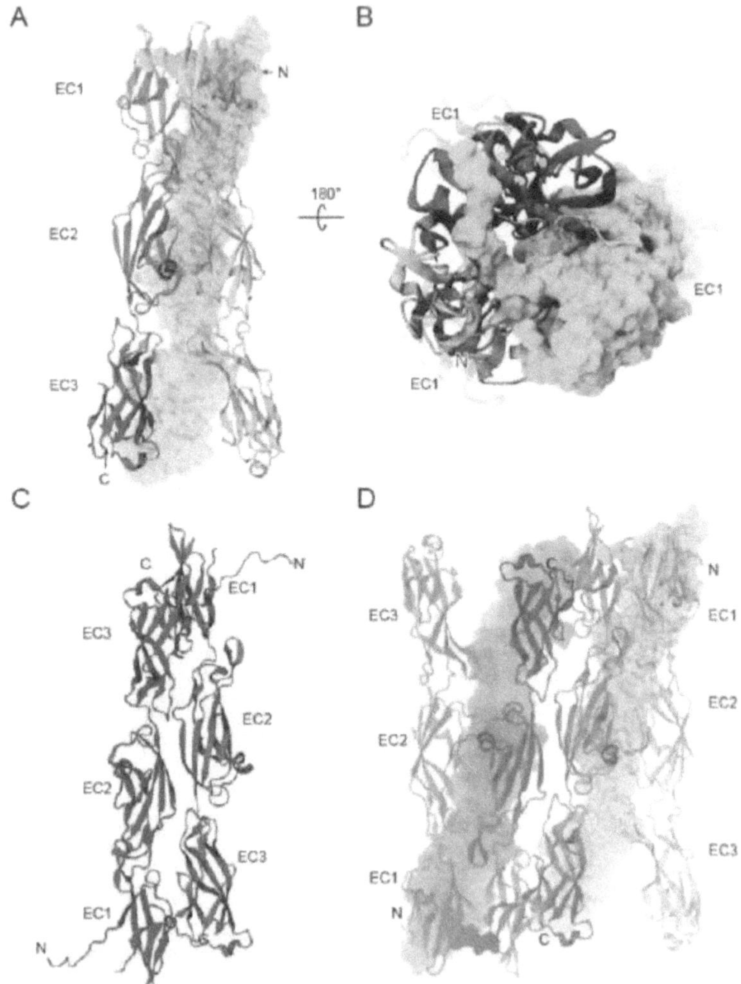

Figure 3.17 Crystal contacts in the *mm* PCDH24 EC1-3 structure.
(A) Crystal contacts show a potential *cis* trimer formed by three parallel monomers. The interface area between two monomers is 1352.6 Å². Two monomers are shown as ribbons while the third one is shown in molecular surface representation. (B) Detail of trimer seen from EC1 (top) shows that the extended N-terminal strands interlace between monomers. (C) Crystal contacts show an antiparallel *trans* dimer with an interface area of 1221.2 Å². (D) Taken together, the potential *cis* and *trans* interfaces form a large complex with two *cis* trimers forming an antiparallel *trans* dimer.

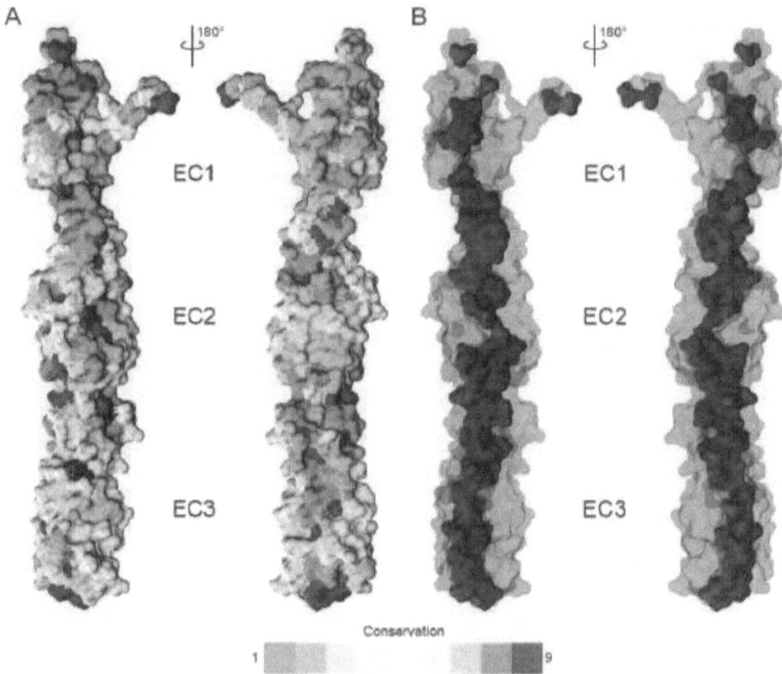

Figure 3.18 Sequence conservation of PCDH24 EC1-3.
(A) Surface representation of *mm* PCDH24 EC1-3 structure with residues colored according to sequence conservation determined using Consurf and a sequence alignment including over 94 species. Teal colors indicate residues that are least conserved while magenta indicates residues that are most conserved among species. (B) Transparent surface representation of *mm* PCDH24 EC1-3 with most conserved residues shown as an opaque magenta surface. Protein core is conserved.

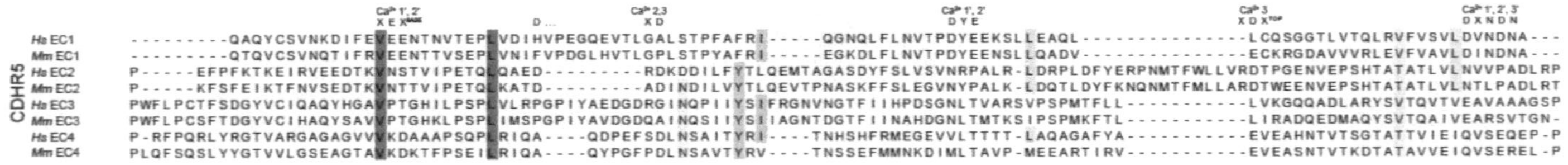

```
                          Ca²⁺ 1',2'                  Ca²⁺ 2,3              Ca²⁺ 1',2'              Ca²⁺ 3              Ca²⁺ 1',2',3'
                          X E X^bulk       D...        X D                  D Y E                   X D X^top           D X N D N
      Hs EC1   ----------QAQYCSVNKDIFEVEENTNVTEPLVDIHVPEGQEVTLGALSTPFAFRI------QGNQLFLNVTPDYEEKSLLEAQL---------------LCQSGGTLVTQLRVFVSVLDVNDNA---
      Mm EC1   ----------QTQVCSVNQTIFRVEENTTVSEPLVNIFVPDGLHVTLGPLSTPYAFRI------EGKDLFLNVTPDYEENSLLQADV----------------ECKRGDAVVVRLEVFVAVLDINDNA---
 CDHR5 Hs EC2  P------EFPFKTKEIRVEEDTKVNSTVIPETQLQAED------------RDKDDILFYTLQEMTAGASDYFSLVSVNRPALR-LDRPLDFYERPNMTFWLLVRDTPGENVEPSHTATATLVLNVVPADLRP
      Mm EC2   P------KFSFEIKTFNVSEDTKVNTTVIPETQLKATD-----------ADINDILVYTLQEVTPNASKFFSLEGVNYPALK-LDQTLDYFKNQNMTFMLLARDTWEENVEPSHTATATLVLNTLPADLRT
      Hs EC3   PWFLPCTFSDGYVCIQAQYHGAVPTGHILPSPLVLRPGPIYAEDGDRGINQPIIYSIFRGNVNGTFIIHPDSGNLTVARSVPSPMTFLL----------LVKGQQADLARYSVTQVTVEAVAAAGSP
      Mm EC3   PWFLPCSFTDGYVCIHAQYSAVVPTGHKLPSPLIMSPGPIYAVDGDQAINQSIIYSIIAGNTDGTFIINAHDGNLTMTKSIPSPMKFTL----------LIRADQEDMAQYSVTQAIVEARSVTGN-
      Hs EC4   P-RFPQRLYRGTVARGAGAGVVVKDAAAPSQPLRIQA----QDPEFSDLNSAITYRI----TNHSHFRMEGEVVLTTTT-LAQAGAFYA----------EVEAHNTVTSGTATTVIEIQVSEQEP-P
      Mm EC4   PLQFSQSLYYGTVVLGSEAGTAVKDKTFPSEILRIQA----QYPGFPDLNSAVTYRV----TNSSEFMMNKDIMLTAVP-MEEARTIRV----------EVEASNTVTKDTATAVVEIQVSEREL-P
```

Figure 3.19 Sequence alignment of hs and mm CDHR5 EC repeats.
All 4 EC repeats for each species are aligned to each other (EC1-4).

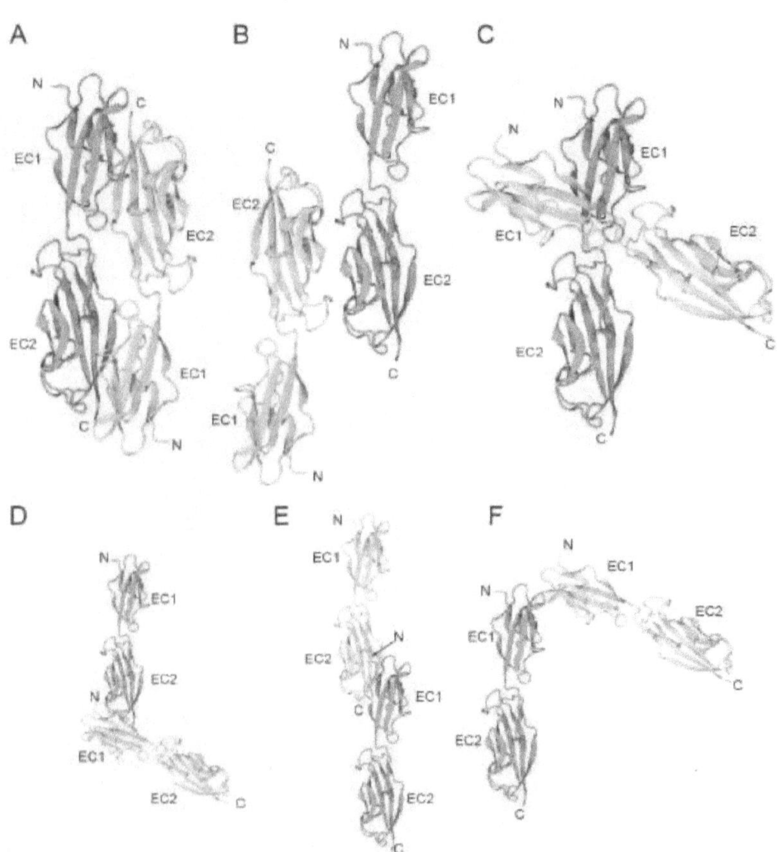

Figure 3.20 Crystal contacts in the *hs* CDHR5 EC1-2 structure.
(A-F) Ribbon diagram of two protomers showing various crystal contacts between monomers of *hs* CDHR5 EC1-2. Interface areas are 1433.9 Å^2 (A), 530.2 Å^2 (B), 364.7 Å^2 (C), 347.9 Å^2 (D), 132.7 Å^2 (E), and 114.0 Å^2 (F), respectively. Interfaces in (A) and (B) correspond to possible *trans* overlaps of EC1-2 and EC1-3, respectively. While *hs* CDHR5 does not mediate *trans* homophilic adhesion based on the homophilic binding assay data presented in Figure 3.4, these interfaces may serve as templates for the mouse CDHR5 protein. Crystal contacts in (D-F) are unlikely to be of physiological relevance due to the arrangement of the monomers.

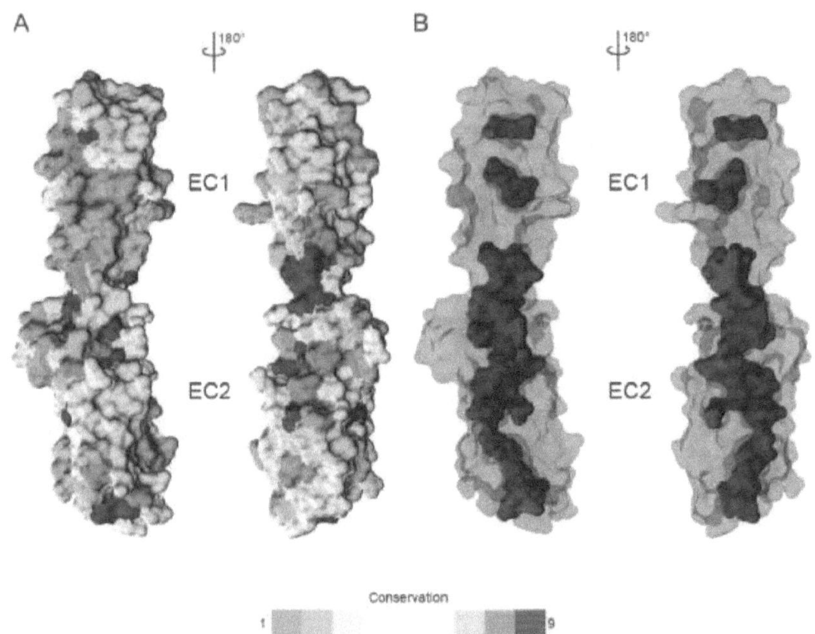

Figure 3.21 Sequence conservation of CDHR5 EC1-2.
(A) Surface representation of *hs* CDHR5 EC1-2 structure with residues colored according to sequence conservation determined using Consurf and a sequence alignment including over 69 species. Teal colors indicate residues that are least conserved while magenta indicates residues that are most conserved among species. (B) Transparent surface representation of *hs* CDHR5 EC1-2 with most conserved residues shown as an opaque magenta surface.

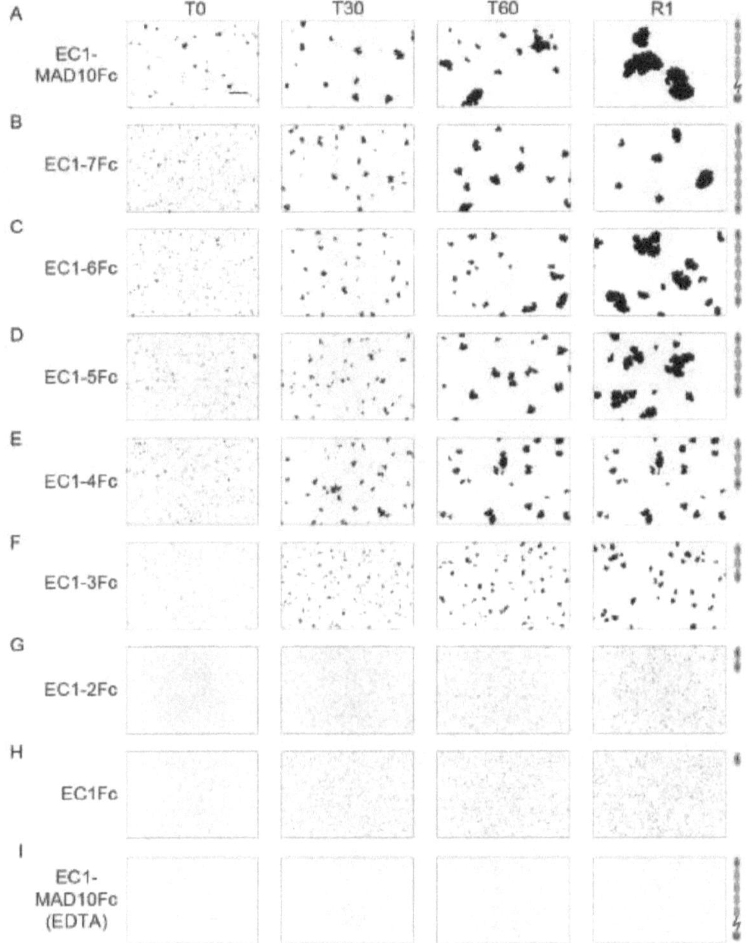

Figure 3.22 Homophilic binding assays of *hs* PCDH24 at various time points.
(A-H) Protein G beads coated with the full-length extracellular domain of *hs* PCDH24 (A) and its C-terminal truncation versions (B-H). Images show bead aggregation observed at the start of the experiment (T0), after 30 min (T30), after 60 min (T60) followed by rocking for 1 min (R1), all in the presence of 2 mM $CaCl_2$. Bars - 500 μm. (I) Protein G beads coated with the full-length extracellular domain of *hs* PCDH24 in the presence of 2 mM EDTA, shown as in (A).

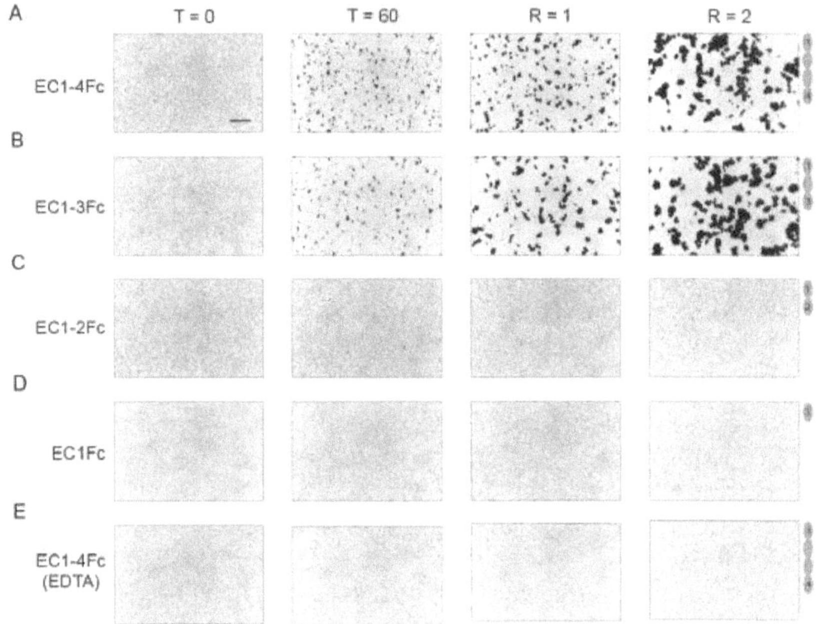

Figure 3.23 Homophilic binding assays of mm CDHR5 at various time points.
(A-D) Protein G beads coated with the full-length cadherin extracellular domain of mm CDHR5 (A) and its C-terminal truncation versions (B-D). Images show the aggregation observed at the start of experiment (T0), after 60 min (T60) followed by rocking for 1 min (R1) and 2 min (R2) in the presence of 2 mM CaCl$_2$. Bars - 500 μm. (E) Protein G beads coated with the full-length cadherin extracellular domain of mm CDHR5 in the presence of 2 mM EDTA shown as in (A).

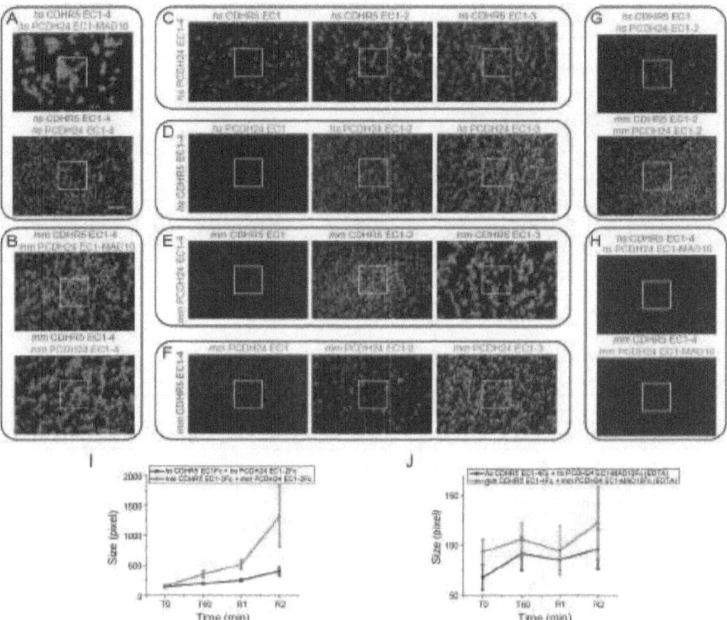

Figure 3.24 Heterophilic binding assays of hs and mm PCDH24 and CDHR5.
(A-B) Images from binding assays of hs PCDH24 (full-length extracellular domain and EC1-4Fc) mixed with the full-length hs CDHR5 cadherin extracellular domain (A), and mm PCDH24 (full-length extracellular domain and EC1-4Fc) mixed with full-length mm CDHR5 cadherin extracellular domain (B). Green fluorescent protein A beads are coated with PCDH24 fragments and red fluorescent protein A beads are coated with CDHR5 in all panels. The white boxes indicate the regions shown in Figure 3.6. Images show bead aggregation observed after 60 min followed by rocking for 2 min in the presence of 2 mM $CaCl_2$. (C-F) Images from binding assays of truncations of hs CDHR5 mixed with hs PCDH24 EC1-4Fc (C), truncations of hs PCDH24 mixed with hs CDHR5 EC1-4Fc (D), truncations of mm CDHR5 mixed with mm PCDH24 EC1-4Fc (E), and truncations of mm PCDH24 mixed with mm CDHR5 EC1-4Fc (F). Images show bead aggregation observed after 60 min followed by rocking for 2 min in the presence of 2 mM $CaCl_2$. (G) Images from binding assays of minimum EC repeats required for heterophilic adhesion of hs and mm PCDH24 and CDHR5. The minimum units for heterophilic adhesion for the human proteins are CDHR5 EC1Fc and PCDH24 EC1-2Fc. The minimum units for heterophilic adhesion for the mouse proteins are CDHR5 EC1-2Fc and PCDH24 EC1-2Fc. Images show bead aggregation observed after 60 minutes followed by rocking for 2 minutes in the presence of 2 mM $CaCl_2$. (H) Protein A beads coated with full-length hs and mm PCDH24 and CDHR5 cadherin extracellular domains (including MAD10 for PCDH24 when indicated and without the mucin-like domain for CDHR5) in the presence of 2 mM EDTA shown as in (A-B). Bar - 500 µm. (I) Aggregate size for minimum units of heterophilic adhesion of hs and mm PCDH24 and CDHR5 at the start of the experiment (T0), after 60 min (T60) followed by rocking for 1 min (R1) and 2 min (R2). (J) Aggregate size for full-length hs and mm PCDH24 and CDHR5 at the start of the experiment (T0), after 60 min (T60) followed by rocking for 1 min (R1) and 2 min (R2). Error bars in I and J are standard error of the mean (n indicated in Table 3.8).

Table 3.2 Percent Identity of EC repeats of CDH23, PCDH15, PCDH24 and CDHR5 proteins across species. The accession numbers of the species used are listed in Tables 3.3-3.6.

CDH23	Repeat % identity	PCDH15	Repeat % identity	PCDH24	Repeat % identity	CDHR5	Repeat % identity
1	64.8	1	59	1	14.2	1	10.3
2	63.5	2	54.2	2	14	2	14.4
3	69.6	3	50	3	20.4	3	16.4
4	62.5	4	39.7	4	22.3	4	2.8
5	50.5	5	44.3	5	19.8	Avg	10.975
6	43.6	6	47.1	6	20.7		
7	67.3	7	37.3	7	15.8		
8	50	8	31.8	8	16.4		
9	44.8	9	39.4	9	9.8		
10	57.9	10	37.6	MAD10	10.9		
11	50.9	11	47.7	Avg	16.43		
12	55.2	MAD12	58.3				
13	43.8	Avg	45.533				
14	36.7						
15	50.5						
16	42.7						
17	51.4						
18	46.3						
19	36.4						
20	34.9						
21	42.9						
22	51.4						
23	53.3						
24	51						
25	52.7						
26	47.7						
27	53						
MAD28	69.2						
Avg	51.589						

Table 3.3 Accession numbers of PCDH24 sequences used for alignment of EC repeats across different species.

Abbreviation	Accession Number	Species	Common Name
Hs	NP_001165447.1	*Homo sapiens*	Human
Pt	XP_016809791.1	*Pan troglodytes*	Chimpanzee
Rn	XP_214434.4	*Rattus norvegicus*	Norwegian rat
Mm	NP_001028536.2	*Mus musculus*	Mouse
Clf	XP_005619238.1	*Canis lupus familiaris*	Dog
Fc	XP_019693959.2	*Felis catus*	Domestic cat
Aj	XP_014928203.2	*Acinonyx jubatus*	Cheetah
Bt	NP_001179233.2	*Bos taurus*	Cow
Ss	XP_013850328.2	*Sus scrofa*	Pig
Oa	XP_004008751.3	*Ovis aries*	Sheep
Pc	XP_020854166.1	*Phascolarctos cinereus*	Koala
Dl	XP_022420063.1	*Delphinapterus leucas*	Beluga whale
Gg	XP_015149366.1	*Gallus gallus*	Chicken
Ccc	XP_019141545.2	*Corvus cornix cornix*	Crow
Af	XP_019329391.1	*Aptenodytes forsteri*	Emperor penguin
Ac	XP_008119531.1	*Anolis carolinensis*	Green anole
Dr	XP_017214654.2	*Danio rerio*	Zebrafish
Am	XP_022532075.1	*Astyanax mexicanus*	Mexican tetra
Om	XP_021417551.1	*Oncorhynchus mykiss*	Rainbow trout

Table 3.4 Accession numbers of CDHR5 sequences used for alignment of EC repeats across different species.

Abbreviation	Accession Number	Species	Common Name
Hs	NP_068743.2	*Homo sapiens*	Human
Pt	XP_016775501.1	*Pan troglodytes*	Chimpanzee
Rn	XP_006230571.1	*Rattus norvegicus*	Norway rat
Mm	NP_001107794.1	*Mus musculus*	Mouse
Fc	XP_019668438.1	*Felis catus*	Domestic cat
Bt	NP_001096796.1	*Bos taurus*	Cow
Ss	XP_013845324.2	*Sus scrofa*	Pig
Oa	XP_027815739.1	*Ovis aries*	Sheep
Pc	XP_020835078.1	*Phascolarctos cinereus*	Koala
Dl	XP_022421100.1	*Delphinapterus leucas*	Beluga whale
Af	XP_009281978.1	*Aptenodytes forsteri*	Emporer penguin
Ac	XP_008106729.1	*Anolis carolinensis*	Green anole
Dr	XP_021326278.1	*Danio rerio*	Zebrafish
Am	XP_022525707.1	*Astyanax mexicanus*	Mexican tetra
Om	XP_021454527.1	*Oncorhynchus mykiss*	Rainbow trout

Table 3.5 Accession numbers of PCDH15 sequences used for EC alignment and conservation analysis.

Accession Number	Species	Common Name
NP_001136235.1	*Homo sapiens*	Human
NP_001136241.1	*Homo sapiens*	Human
XP_019792025.1	*Tursiops truncatus*	Bottlenose dolphin
XP_020929194.1	*Sus scrofa*	pig
XP_022266127.1	*Canis lupus familiaris*	Dog
NP_075604.2	*Mus musculus*	mouse
XP_024416857.1	*Desmodus rotundus*	Common vampire bat
XP_015143564.1	*Gallus gallus*	Chicken
XP_021147002.1	*Columbia livia*	Pigeon
XP_021391314.1	*Lonchura striata domestica*	Finch
XP_017589248.1	*Corvus brachyrhynchos*	American Crow
XP_020662251.1	*Pogona vitticeps*	Bearded dragon
XP_016851436.1	*Anolis carolinensis*	Green anole
XP_019394135.1	*Crocodylus porosus*	Australian crocodile
NP_001012500.1	*Danio rerio*	Zebrafish
XP_012675462.1	*Clupea harengus*	Atlantic herring
XP_022525074.1	*Astyanax mexicanus*	Mexican tetra
XP_007897895.1	*Callorhinchus milii*	Australian ghostshark

Table 3.6 Accession numbers of CDH23 sequences used for EC alignment and conservation analysis.

Accession Number	Species	Common Name
NP_071407.4	*Homo sapiens*	Human
XP_016774005.2	*Pan troglodytes*	Chimpanzee
XP_001925718.2	*Sus scrofa*	Pig
NP_001178135.2	*Bos taurus*	Cow
XP_014695776.1	*Equus asinus*	Donkey
XP_023096304.1	*Felis catus*	Cat
XP_022273252.1	*Canis lupus familiaris*	Dog
XP_007454751.1	*Lipotes vexillifer*	Yangtze river dolphin
NP_446096.1	*Rattus norvegicus*	Norwegian rat
NP_075859.2	*Mus musculus*	Mouse
XP_013028252.1	*Anser cygnoides domesticus*	Domestic goose
XP_015143732.1	*Gallus gallus*	Chicken
XP_015278377.1	*Gekko japonicus*	Japanese gekko
XP_019339897.1	*Alligator mississippiensis*	American alligator
XP_007897937.1	*Callorhinchus milii*	Australian ghostshark
NP_999974.1	*Danio rerio*	Zebrafish
XP_012672424.1	*Clupea harengus*	Atlantic herring

Table 3.7 Accession numbers of sequences of PCDH24, CDHR5, CDH23, PCDH15, CDH1, and CDH2 used for generating Figure 1B.

Species	Accession Number for EC1-3					
	PCDH24	CDHR5	CDH23	PCDH15	CDH1	CDH2
Homo sapiens	NP_001165447.1	NP_068743.3	NP_071407.4	NP_001136235.1	NP_004351.1	NP_001783.2
Mus Musculus	NP_001028536.2	NP_001107794.1	NP_075859.2	NP_075604.2	NP_033994.1	NP_031690.3
Danio rerio	XP_017214654.2	XP_021326278.1	NP_999974.1	NP_001012500.1	NP_571895.1	NP_571156.2
Anolis carolensis	XP_008119531.1	XP_008106723.1	XP_016847668.1	XP_016851436.1	XP_008121673.2*	XP_008106822.1
Gallus gallus	XP_015149366.1	NA	XP_421595.4	NP_001038119.1	NP_001034347.2	NP_001001615.1

* indicates a low-quality sequence

Table 3.8 Number of biological replicates for bead aggregation assays.

Species	Protein		Timepoints	n
Human	PCDH24	EC1	T = 0, 30, 60; R = 1	4
		EC1-2	T = 0, 30, 60; R = 1	4
		EC1-3	T = 0, 30, 60; R = 1	3
		EC1-4	T = 0, 30, 60; R = 1	3
		EC1-5	T = 0, 30, 60; R = 1	3
		EC1-6	T = 0, 30, 60; R = 1	3
		EC1-7	T = 0, 30, 60; R = 1	3
		EC1-MAD10	T = 0, 30, 60; R = 1	4
		EC1-MAD10*	T = 0, 60; R = 1	3
Mouse	PCDH24	EC1-MAD10	T = 0, 60; R = 1	3
Human	CDHR5	EC1-4	T = 0, 60; R = 1, 2	3
Mouse	CDHR5	EC1	T = 0, 60; R = 1, 2	3
		EC1-2	T = 0, 60; R = 1, 2	3
		EC1-3	T = 0, 60; R = 1, 2	3
		EC1-4	T = 0, 60; R = 1, 2	3
		EC1-4 E84G	T = 0, 60; R = 1, 2	3
		EC1-4 R82G	T = 0, 60; R = 1, 2	3
Human	CDHR5 EC1 + PCDH24 EC1-4		T = 0, 60; R = 1, 2	4
	CDHR5 EC1-2 + PCDH24 EC1-4		T = 0, 60; R = 1, 2	4
	CDHR5 EC1-3 + PCDH24 EC1-4		T = 0, 60; R = 1, 2	4
	CDHR5 EC1-4 + PCDH24 EC1-4		T = 0, 60; R = 1, 2	5
	CDHR5 EC1-4 + PCDH24 EC1		T = 0, 60; R = 1, 2	3
	CDHR5 EC1-4 + PCDH24 EC1-2		T = 0, 60; R = 1, 2	3
	CDHR5 EC1-4 + PCDH24 EC1-3		T = 0, 60; R = 1, 2	3
	CDHR5 EC1-4 + PCDH24 EC1-MAD10		T = 0, 60; R = 1, 2	3
	CDHR5 EC1 + PCDH24 EC1-2		T = 0, 60; R = 1, 2	3
Mouse	CDHR5 EC1 + PCDH24 EC1-4		T = 0, 60; R = 1, 2	3
	CDHR5 EC1-2 + PCDH24 EC1-4		T = 0, 60; R = 1, 2	3
	CDHR5 EC1-3 + PCDH24 EC1-4		T = 0, 60; R = 1, 2	3
	CDHR5 EC1-4 + PCDH24 EC1-4		T = 0, 60; R = 1, 2	3
	CDHR5 EC1-4 + PCDH24 EC1		T = 0, 60; R = 1, 2	3
	CDHR5 EC1-4 + PCDH24 EC1-2		T = 0, 60; R = 1, 2	3
	CDHR5 EC1-4 + PCDH24 EC1-3		T = 0, 60; R = 1, 2	3
	CDHR5 EC1-4 + PCDH24 EC1-MAD10		T = 0, 60; R = 1, 2	3
	CDHR5 EC1-2 + PCDH24 EC1-2		T = 0, 60; R = 1, 2	3

* Used for the *mm* PCDH24 EC1-MAD10 as a control

Table 3.9 Accession numbers of sequences of PCDH24 used in Consurf.

Accession Number	Species	Common Name
NP_001165447.1	*Homo sapiens*	Human
XP_014928203.2	*Acinonyx jubatus*	Cheetah
NP_001179233.2	*Bos taurus*	Cow
XP_005619238.1	*Canis lupus familiaris*	Dog
XP_022420063.1	*Delphinapterus leucas*	Beluga whale
XP_019693959.2	*Felis catus*	Domestic cat
NP_001028536.2	*Mus musculus*	Mouse
XP_004008751.3	*Ovis aries*	Sheep
XP_016809791.1	*Pan troglodytes*	Chimpanzee
XP_214434.4	*Rattus norvegicus*	Norwegian rat
XP_013850328.2	*Sus scrofa*	Pig
XP_014996966.2	*Macaca mulatta*	Rhesus macaque
XP_015307761.1	*Macaca fascicularis*	Crab-eating macaque
XP_012974778.1	*Mesocricetus auratus*	Golden hamster
XP_004737588.1	*Mustela putorius furo*	Domestic ferret
XP_023472938.1	*Equus caballus*	Horse
XP_029075003.1	*Monodon monoceros*	Narwhal
XP_028908976.1	*Ornithorhynchus anatinus*	Platypus
XP_008253668.2	*Oryctolagus cuniculus*	Rabbit
XP_008988594.1	*Callithrix jacchus*	White-tufted-ear marmoset
XP_024102684.1	*Pongo abelii*	Sumatran orangutan
XP_003473404.1	*Cavia porcellus*	Domestic guinea pig
XP_027263978.1	*Cricetulus griseus*	Chinese hamster
XP_023371131.1	*Otolemur garnettii*	Small-eared galago
XP_003806887.1	*Pan paniscus*	Pygmy chimpanzee
XP_003900612.1	*Papio anubis*	Olive baboon
XP_018883422.2	*Gorilla gorilla gorilla*	Western lowland gorilla
XP_004284878.1	*Orcinus orca*	Killer Whale
XP_019798890.1	*Tursiops truncatus*	Common bottlenose dolphin
XP_019329391.1	*Aptenodytes forsteri*	Emperor penguin
XP_019141545.2	*Corvus cornix cornix*	Crow
XP_015149366.1	*Gallus gallus*	Chicken
XP_027752218.1	*Empidonax traillii*	Willow Flycatcher
XP_027574630.1	*Pipra filicauda*	Wire-tailed manakin
XP_027538397.1	*Neopelma chrysocephalum*	Saffron-crested tyrant-manakin
XP_027499547.1	*Corapipo altera*	White-ruffed manakin
XP_026714386.1	*Athene cunicularia*	Burrowing owl
XP_025978389.1	*Dromaius novaehollandiae*	Emu

XP_025903300.1	*Nothoprocta perdicaria*	Chilean tinamou
XP_023791618.1	*Cyanistes caeruleus*	Blue tit
XP_021266449.1	*Numida meleagris*	Helmeted guineafowl
XP_017683677.1	*Lepidothrix coronata*	Blue-crowned manakin
XP_015731588.1	*Coturnix japonica*	Japanese quail
XP_015496956.1	*Parus major*	Great tit
XP_014803238.1	*Calidris pugnax*	Ruff
XP_013797945.1	*Apteryx australis mantelli*	North Island brown kiwi
XP_011569753.1	*Aquila chrysaetos canadensis*	American golden eagle
XP_019475977.1	*Meleagris gallopavo*	Turkey
XP_005053635.1	*Ficedula albicollis*	Collared flycatcher
XP_027634673.1	*Falco peregrinus*	Peregrine falcon
XP_027657949.1	*Falco cherrug*	Saker falcon
XP_026647227.1	*Zonotrichia albicollis*	White-throated sparrow
XP_021144536.1	*Columba livia*	Rock pigeon
XP_014108468.1	*Pseudopodoces humilis*	Tibetan ground-tit
XP_017583659.1	*Corvus brachyrhynchos*	American crow
XP_008939384.1	*Merops nubicus*	Carmine bee-eater
XP_009089922.2	*Serinus canaria*	Common canary
XP_009331800.1	*Pygoscelis adeliae*	Adelie penguin
XP_008119531.1	*Anolis carolinensis*	Green anole
XP_026576910.1	*Pseudonaja textilis*	Eastern brown snake
XP_026540941.1	*Notechis scutatus*	Mainland tiger snake
XP_020646672.1	*Pogona vitticeps*	Central bearded dragon
XP_019390868.1	*Crocodylus porosus*	Australian saltwater crocodile
XP_013925415.1	*Thamnophis sirtalis*	Common garter snake
XP_022532075.1	*Astyanax mexicanus*	Mexican tetra
XP_017214654.2	*Danio rerio*	Zebrafish
XP_021417551.1	*Oncorhynchus mykiss*	Rainbow trout
XP_014053985.1	*Salmo salar*	Atlantic salmon
XP_007904435.1	*Callorhinchus milii*	Elephant shark
XP_024139064.1	*Oryzias melastigma*	Indian Medaka
XP_022607721.1	*Seriola dumerili*	Greater amberjack
XP_015832622.1	*Nothobranchius furzeri*	Turquoise killifish
XP_004073569.1	*Oryzias latipes*	Japanese medaka
XP_029021466.1	*Betta splendens*	Siamese fighting fish
XP_028315581.1	*Gouania willdenowi*	Blunt-snouted clingfish

XP_028271476.1	*Parambassis ranga*	Indian glassy fish
XP_027865185.1	*Xiphophorus couchianus*	Monterrey platyfish
XP_026993994.1	*Tachysurus fulvidraco*	Yellow catfish
XP_026870202.1	*Electrophorus electricus*	Electric eel
XP_026802869.1	*Pangasianodon hypophthalmus*	Striped catfish
XP_026214145.1	*Anabas testudineus*	Climbing perch
XP_026186036.1	*Mastacembelus armatus*	Zig-zag eel
XP_026045273.1	*Astatotilapia calliptera*	Eastern happy
XP_019200886.1	*Oreochromis niloticus*	Nile tilapia
XP_014326181.1	*Xiphophorus maculatus*	Southern platyfish
XP_014187335.1	*Haplochromis burtoni*	Burton's mouthbrooder
XP_014342354.1	*Latimeria chalumnae*	Coelacanth
XP_007561596.1	*Poecilia formosa*	Amazon molly
XP_024918834.1	*Cynoglossus semilaevis*	Tongue sole
XP_010735796.2	*Larimichthys crocea*	Large yellow croaker
XP_010898483.1	*Esox lucius*	Northern pike
XP_012724017.1	*Fundulus heteroclitus*	Mummichog
XP_013870118.1	*Austrofundulus limnaeus*	Killifish
XP_014849149.1	*Poecilia mexicana*	Atlantic molly

Table 3.10 Accession numbers of sequences of CDHR5 used in Consurf.

Accession Number	Species	Common Name
NP_068743.2	*Homo sapiens*	Human
NP_001107794.1	*Mus musculus*	Mouse
XP_027815739.1	*Ovis aries*	Sheep
XP_022421100.1	*Delphinapterus leucas*	Beluga whale
XP_020835078.1	*Phascolarctos cinereus*	Koala
XP_016775501.1	*Pan troglodytes*	Chimpanzee
XP_006230571.1	*Rattus norvegicus*	Norway rat
XP_013845324.2	*Sus scrofa*	Pig
NP_001096796.1	*Bos taurus*	Cow
XP_019668438.1	*Felis catus*	Domestic cat
XP_023510521.1	*Equus caballus*	Horse
XP_014437989.1	*Tupaia chinensis*	Chinese tree shrew
XP_012973076.1	*Mesocricetus auratus*	Golden hamster
XP_029064308.1	*Monodon monoceros*	Narwhal
XP_028642741.1	*Grammomys surdaster*	African woodland thicket rat
XP_028370928.1	*Phyllostomus discolor*	Pale spear-nosed bat
XP_025847484.1	*Vulpes vulpes*	Red fox
XP_025717837.1	*Callorhinus ursinus*	Northern ful seal
XP_024430321.1	*Desmodus rotundus*	Common vampire bat
XP_020767741.1	*Odocoileus virginianus texanus*	White-tailed deer
XP_009281978.1	*Aptenodytes forsteri*	Emporer penguin
XP_025966497.1	*Dromaius novaehollandiae*	Emu
XP_025928961.1	*Apteryx rowi*	Okarito brown kiwi
XP_025900649.1	*Nothoprocta perdicaria*	Chilean tinamou
XP_023784276.1	*Cyanistes caeruleus*	Blue tit
XP_015485075.1	*Parus major*	Great tit
XP_015718593.1	*Coturnix japonica*	Japanese quail
XP_014796130.1	*Calidris pugnax*	Ruff
XP_014734396.1	*Sturnus vulgaris*	Common starling
XP_013800822.1	*Apteryx australis mantelli*	North Island brown kiwi
XP_013053829.1	*Anser cygnoides domesticus*	Domestic goose
XP_011595784.1	*Aquila chrysaetos canadensis*	American golden eagle
XP_010709142.1	*Meleagris gallopavo*	Turkey
XP_010562687.1	*Haliaeetus leucocephalus*	Bald eagle
XP_010297449.1	*Balearica regulorum gibbericeps*	East African grey crowned-crane
XP_010292632.1	*Phaethon lepturus*	White-tailed tropicbird
XP_010225820.1	*Tinamus guttatus*	White-throated tinamou
XP_010207507.1	*Colius striatus*	Speckled mousebird

XP_007907548.1	*Callorhinchus milii*	Elephant shark
XP_021454527.1	*Oncorhynchus mykiss*	Rainbow trout
XP_021326278.1	*Danio rerio*	Zebrafish
XP_022525707.1	*Astyanax mexicanus*	Mexican tetra
XP_024120964.1	*Oryzias melastigma*	Indian medaka
XP_023807528.1	*Oryzias latipes*	Japanese medaka
XP_029312546.1	*Cottoperca gobio*	Channel bull blenny
XP_027008843.1	*Tachysurus fulvidraco*	Yellow catfish
XP_026168289.1	*Mastacembelus armatus*	Zig-zag eel
XP_019112627.2	*Larimichthys crocea*	Large yello croaker
XP_017341402.1	*Ictalurus punctatus*	Channel catfish
XP_016392038.1	*Sinocyclocheilus rhinocerous*	Rayfinned fish
XP_016324417.1	*Sinocyclocheilus anshuiensis*	Rayfinned fish
XP_015227070.1	*Cyprinodon variegatus*	Sheepshead minnow
XP_021163353.1	*Fundulus heteroclitus*	Mummichog
XP_012674652.1	*Clupea harengus*	Atlantic herring
XP_005938389.1	*Haplochromis burtoni*	Burton's mouthbrooder
XP_004556350.2	*Maylandia zebra*	Zebra mbuna
XP_008106729.1	*Anolis carolinensis*	Green anole
XP_019340716.1	*Alligator mississippiensis*	American alligator
XP_028591121.1	*Podarcis muralis*	Common wall lizard
XP_026562223.1	*Pseudonaja textilis*	Eastern brown snake
XP_026523924.1	*Notechis scutatus*	Mainland tiger snake
XP_026513403.1	*Terrapene carolina triunguis*	Three-toed box turtle
XP_019402850.1	*Crocodylus porosus*	Australian saltwater crocodile
XP_019375347.1	*Gavialis gangeticus*	Gharial
XP_015745583.1	*Python bivittatus*	Burmese python
XP_013925033.1	*Thamnophis sirtalis*	Common garter snake
XP_014429985.1	*Pelodiscus sinensis*	Chinese softshell turtle
XP_025054327.1	*Alligator sinensis*	Chinese alligator
XP_023967188.1	*Chrysemys picta bellii*	Western painted turtle

Table 3.11 Summary of simulations

Label	System	Type	Start	Time (ns)	Pulling speed (nm/ns)	Average force peak (pN)[d]	Size of system (# atoms)	Box dimensions (Å³)
S1a	Hs PCDH24 EC1-2	Equil[a]	-	10.8	-	-	125,348	227 × 88 × 67
S1b		SMD[b]	S1a	1	10	675.2		
S1c		SMD[b]	S1a	21	1	383.9		
S1d		SMD[b]	S1a	58.6	0.1	167.1		
S2	Mm PCDH24 EC1-3	Equil[a]	-	99.1	-	-	227,299	171 × 112 × 111
S3a	Mm PCDH24 EC1-3	Equil[a]	-	11.1	-	-	195,074	337 × 71 × 80
S3b		SMD[c]	S3a	1	10	1905.5		
S3c		SMD[c]	S3a	10	1	1385.4		
S3d		SMD[c]	S3a	70	0.1	1216.0		

[a] Indicates simulations that consisted of 1,000 steps of minimization, 100 ps of dynamics with the protein backbone constrained, 1 ns of free dynamics in the NpT ensemble ($\gamma = 1$ ps^{-1}), and then of free dynamics in the NpT ensemble ($\gamma = 0.1$ ps^{-1}) for the rest of the time.

[b] Indicates SMD simulations in which the force was applied to the Cα atoms of residues p.P215 of chains A and D.

[c] Indicates SMD simulations in which the force was applied to the centers of mass (COM) of EC1 (p.N1 to p.E95) and EC3 (p.P215 and p.T324).

[d] Average peak forces were calculated from peak forces applied to stretched Cα atoms or COM at each end and using 50 ps running averages.

Chapter 4. Conclusions

The cadherin superfamily is formed by a diverse group of calcium-dependent cell adhesion proteins, with a wide variety of adhesive interfaces that allow them to play roles in embryogenesis, tissue morphogenesis and maintenance, brain wiring, and mechanotransduction[8,27,33,109,189]. The three subfamilies, classical cadherins, clustered protocadherins and non-clustered protocadherins, are defined by how they mediate adhesion[7]. The non-clustered protocadherins are by far the most diverse to date. In this book, we explored the adhesion mechanisms of several less studied non-clustered protocadherins: cadherin 17 (CDH17), cadherin 16 (CDH16), protocadherin 24 (PCDH24), mucin-like Protocadherin (CDHR5), and protocadherin 21 (PCH21). Each of these cadherin family members performs a unique function within the body and therefore each relies on different *trans* and *cis* interactions to mediate cell adhesion.

CDH17 and CDH16 are defined as being related to classical cadherins, but have two additional N-terminal EC repeats, lack a tryptophan residue in their EC1 repeat to mediate strand-swapping, and have short cytoplasmic tails[77,93,218]. They are expressed in systems primarily designed for water and nutrient absorption and waste removal. CDH17 has been hypothesized to play a role in regulating water transport between cells[71,72], and both CDH17 and CDH16 are involved in a variety of cancers in their respective systems[91,219]. PCDH24 and CDHR5 have been shown to form the essential intermicrovillar links within the brush border epithelia through *in vitro* and *in vivo* assays, including knockdowns and knockouts[56,67,100,103]. These proteins help maintain the organization and uniformity of the microvilli and disruption of them leads to disorder of the brush border[67]. Their cytoplasmic binding partners have been implicated in disease[56,101,104] and a mutation known to cause deafness in the related protocadherin-15 (PCDH15)[66] was replicated in CDHR5 to demonstrate that these related family members might use a similar mechanism of adhesion to the tip-link protocadherins involved in hearing[67]. Protocadherin-21 (PCDH21) is related to PCDH24 and cadherin-23 (CDH23)[7,109], and it is responsible for the proper formation of the outer segment discs in the photoreceptors within the retina[113]. Mutations in PCDH21 lead to progressive blindness due to changes in the formation of the outer segment and loss of discs[113,115]. All of these cadherins have been studied through various biophysical and cell biology techniques, however, their mechanisms of adhesion have only been hypothesized based on their relationships to other cadherins and therefore remain unclear. In this work, we have used a combination of X-ray crystallography, bead aggregation assays, sequence analysis and molecular dynamics (MD) simulations to further shed light on how each of these cadherins might mediate adhesion within their respective systems.

In Chapter 2, we describe how we were able to determine a structure for CDH17 EC1-2 from X-ray crystallography. Additionally, we were able to determine via bead aggregation assays that EC7 is essential for CDH17 to mediate *trans* adhesion.

Additionally we determined that EC1 is non-essential for CDH17 to mediate *trans* adhesion however the mutant p.L77G, which is present in EC1, abolishes aggregation in bead aggregation assays[72]. We presented mutations in residues that may play key roles in CDH17 function and obtained mixed results, as the p.N154S mutation, involved in disease[83] does not abolish bead aggregation, while both mutations of p.W217, assumed to be similar to the classical cadherin p.W2, did not aggregate. This is the first instance of a cadherin requiring both its most C-terminal repeat to mediate *trans* adhesion and we were able to suggest several potential interfaces based on the crystal structure and bead aggregation assay data as well as in comparison to the interface observed for EC1-4[151]. Sequence analysis demonstrated that CDH17 is not well conserved across all species and that this mechanism might only be used by CDH17 from mammalian species. However, the data presented in Chapter 2 is insufficient and we can only speculate on the repeats involved in the CDH17 *trans* interface.

To that end, we have begun several experiments which may help us elucidate the structure of the *trans* adhesion interface. The full-length extracellular domain of CDH17, consisting of all seven EC repeats has been expressed and purified from Expi293 cells, a line of suspension HEK293 cells that are able to robustly produce secreted proteins. In the case of CDH17, we are able to obtain clean peaks for the entire extracellular domain from his-tag purification and size-exclusion chromatography (Figure 4.1A). The CDH17 EC1-7 fragment produced with glycosylation runs as a broad peak in the Superdex S200 10/300, likely due to heterogenous glycosylation. When an inhibitor, kifunensine, is used to inhibit the longer branches of glycosylation, two peaks in the size-exclusion chromatography trace can be observed. Non-reduced SDS-PAGE reveals that the peak corresponding to a higher molecular weight, possibly a dimer, is not due to intermolecular disulfide bond formation. Both the glycosylated and un-glycosylated proteins were used for crystallographic screens. Unglycosylated protein crystals did not diffract well enough to obtain diffraction datasets for structure determination. However, the glycosylated protein showed a preference for crystallization in buffers containing NaI or NaBr, and PEG 3350 and several low-resolution datasets (Figure 4.1B) have been obtained from a variety of crystals from refinements of those original screen buffers. Initial data processing for these low-resolution sets has been difficult and will likely require several smaller structural models in order to be solved using molecular replacement. In addition, given that we can robustly produce mammalian expressed CDH17 EC1-7, we are planning to use other biophysical techniques such as small-angle X-ray scattering (SAXS), analytical ultracentrifugation (AUC), and multi-angle light scattering (MALS) to find a low-resolution envelope structure of the dimer and to determine binding affinity (K_D) for the protein homodimer. Electron cryo-microscopy (Cryo-EM) is another possibility but would rely on the formation of larger oligomers, as cadherins are quite thin in comparison to globular and membrane bound proteins.

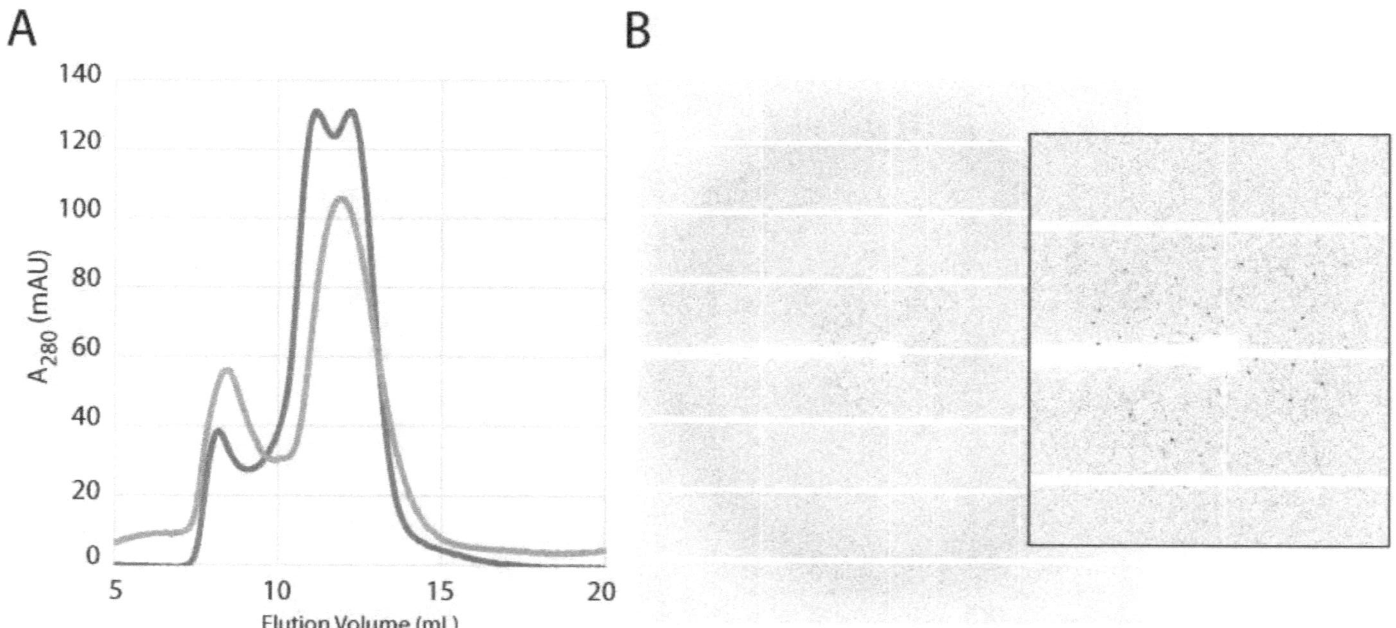

Figure 4.1 Purification and diffraction of CDH17 EC1-7
A) Size-exclusion chromatography traces shown are for CDH17 EC1-7 with (blue) and without (orange) glycosylation inhibitor obtained using the Superdex S200 10/300 size-exclusion column. B) Diffraction pattern of a crystal of CDH17 EC1-7 obtained from beamline 24-ID-E at the Advanced Photo Source (APS).

Additionally, we plan to continue using the bead aggregation assays to further determine the homophilic interface through thoughtfully planned mutations. In order to determine mutations that might affect the *trans* interface, we can use EVCouplings[220–225], a software package that compares protein sequences to similar proteins within the family and across different species and looks for sequence covariance. Covariance can happen within individual domains or between different domains. Here, we are interested in covariance between different EC repeats in CDH17. For example, we may look for covariance between residues EC2 and EC7, and EC3 and EC6, in the case that CDH17's extracellular domain overlaps in a linear fashion. Residues that are coupled could be mutated to interfere with the opposing interface, either by adding bulky hydrophobic groups or opposite charges to disrupt potential salt bridges. These mutations could be used in bead aggregation assays with mutant proteins as a qualitative way to determine which mutations interfere with aggregation. These mutated proteins could then be used with the other biophysical experiments to quantify how the mutation affects homophilic binding.

Thus far, our experiments have focused on CDH17 binding in the presence of excess calcium and likely all CDH17 calcium-binding sites are occupied. Interestingly, experiments by others have suggested that CDH17 is sensitive to changes in calcium concentration[71,72,89] in the interstitial cleft between epithelial cells and is responsible for modulating cleft width as osmotic conditions fluctuate in the intestines[71,72]. Understanding how CDH17 mediates adhesion in the presence of different calcium concentrations could further shed light onto its biological role *in vivo*. Perhaps residues on the surface are able to interact with the weakest bound calcium ion (typically site 1 and exposed) of another protomer. If the site is unoccupied, it could weaken adhesion or modify the way in which CDH17 binds to itself. This would need to be explored in a more quantitative manner than what the bead aggregation assays could do, with additional complications from a likely heterogenous sample. Cryo-EM could be useful, if dimers of CDH17 could be frozen at different calcium concentrations, to observe snapshots of how binding changes.

CDH17 is also implicated in cancer, and its aberrant expression could be a good target for cancer therapeutics. Cancer drugs that bind to specific membrane-bound proteins on specific cancer cells gives them specificity, thus avoiding healthy tissue and causing less side effects. Because we now have a structure of the N-terminal EC repeats of CDH17, we might be able to design a protein-based targeted therapeutic that can bind to EC1 or EC2 of CDH17. Antibody-based targeted therapeutics have already been investigated for CDH17, but antibodies sometimes affect other targets if these are not monoclonal. Epithelial cells produce many different cadherins on their surface, which might be affected by polyclonal antibodies. Listeria monocytogenes produces a protein called internalin, which binds to CDH1 and allows the bacteria to infect epithelia cells[226,227]. A structure of *L. monocytogenes* internalin bound to EC1 of CDH1 exists[227,228]. It contains leucine-rich repeats that curve around the EC repeat, preventing *trans* adhesion from occurring and signaling for endocytosis to begin[226]. If the internalin protein could be mutated in such a way that it binds to CDH17 EC1, it may be able to target CDH17-expressing cancer cells, delivering a drug to kill them.

The experiments described above will help us further understand what the binding interface is for CDH17 and finally shed light on how the 7D-cadherins mediate adhesion. Additionally, experiments on CDH16 can confirm if the 7D-cadherins form the same unique interface, given its overall similarity to CDH17[93]. Like CDH17, CDH16 lacks a tryptophan in EC1 to mediate strand-swapping and also similar atypical calcium-binding motifs in EC2. Both CDH17 and CDH16 are expressed in cells that are involved in water and waste transport and absorption. However, the biological role of CDH16 has not been investigated. It might be possible for CDH16 to modulate the cleft width in the kidney epithelia, similar to how CDH17 does within intestinal epithelia[71,72].

In Chapter 3, we reported how PCDH24 and CDHR5 form the intermicrovillar tip-links. We produced the first structural models for *hs* PCDH24 EC1-2, *mm* PCDH24 EC1-3, and *hs* CDHR5 EC1-2 using X-ray crystallography. The *hs* PCDH24 EC1-2 and *hs* CDHR5 EC1-2 structures showed remarkable similarity to their Cr-2 and Cr-3 siblings, CDH23 and PCDH15, while *mm* PCHD24, at its N-terminus, appeared similar to classical cadherins. Their crystallographic packing showed a variety of homophilic interfaces, which we could further investigate using steered molecular dynamics (SMD) simulations[214], as well as bead aggregation assays. Bead aggregation assays performed on PCDH24 revealed that as expected, *hs* PCDH24 mediates homophilic adhesion[67]. Uniquely, we were able to observe aggregation with only EC1 or EC1-2, though bead aggregates were smaller than what was observed for EC1-3 repeats. This was supported by SMD simulations predicting that a *trans* EC1-2 interface in the *hs* PCDH24 EC1-2 structure was mechanically weak. Unexpectedly, bead aggregation assays performed on *mm* PCDH24 demonstrated that it does not mediate a *trans* homophilic interaction. The opposite is true, however, for CDHR5. Mouse CDHR5 is capable of mediating a homophilic *trans* interaction using its EC1-3 repeats, while, based on earlier research and our own experiments, *hs* CDHR5 is unable to mediate such an interaction.

Our experiments do not show the structure of a heterophilic interface for either human or mouse intermicrovillar links. This is due to difficulty with co-crystallization. In the case of *hs* PCDH24 and *hs* CDHR5, we are able to produce and mix both proteins, but the resulting crystals result in datasets corresponding to CDHR5 only. We have begun to express fragments of PCDH24 and CDHR5 in mammalian cells assuming that longer and glycosylated constructs will form crystals that contain both proteins. However, there are computational methods we can use to determine residues involved in the interface. Because we have structures of both PCDH24 EC1-2 and CDHR5 EC1-2, we can submit those structures to docking servers which output dimeric structures based on interaction energies. These methods are limited in that they are typically used for protein and ligand interactions, rather than with two proteins. We can additionally use evolutionary methods, such as EV-couplings[220–225] mentioned above. However, the low sequence conservation for PCDH24 and CDHR5 could make this difficult. Because these proteins use different EC repeats to mediate adhesion, the residues involved in the interface may not necessarily be the most well-conserved or evolutionarily coupled.

The differences in heterophilic and homophilic binding, as well as the low sequence conservation of PCDH24 and CDHR5, pose interesting biological questions about the evolution of cadherins. Humans and animals all have a range of diets and the layout of organs in the digestive tract is built to support these individual diets. Additionally, microbiomes in the digestive tract are unique amongst different species. Is it possible that the molecular mechanism of adhesion between the intermicrovillar cadherins evolved alongside the human or animals' dietary needs? And is this difference only reserved for the extracellular portions, or do the cytoplasmic domains of PCDH24 and CDHR5 also use unique mechanisms to bind to different partners? The evolution of the cytoplasmic binding partners for cadherins, catenins, existed before cadherins did but their function has remained similar over evolution: directing polarization of cells[203,229]. We know that PCDH24 and CDHR5 use binding partners in the cytoplasm that are the same for the tip-link protocadherins, CDH23 and PCDH15, which are highly conserved across animals[33]. The structures of the inner ear differ greatly between species and simulations of CDH23 and PCDH15 from various species, as well as of ancestral reconstructed proteins reveal that tip-link bonds have different mechanical strengths. The small intestine is more complex than the inner ear because it interacts with a wide variety of microorganisms that belong to the microbiome and responds to changing salt concentrations, osmotic pressure, and nutrient concentrations. We have just begun to open the door to many questions about how the components of the intermicrovillar adhesion complex (IMAC) has evolved as species diets changed and adapted to the developing planet.

In conclusion, the work done in this book has demonstrated that non-clustered, non-classical cadherins utilize diverse mechanisms in order to adhere cells to one another. My research has only focused on a small subset of these unique cadherins and other related family members' adhesive mechanisms remain to be determined.

www.ingramcontent.com/pod-product-compliance
Lightning Source LLC
LaVergne TN
LVHW042149190726
843493LV00006B/1590